Die Grundlehren der mathematischen Wissenschaften

in Einzeldarstellungen
mit besonderer Berücksichtigung
der Anwendungsgebiete

Band 135

Handbook for Automatic Computation

Edited by

F. L. Bauer · A. S. Householder · F. W. J. Olver
H. Rutishauser · K. Samelson · E. Stiefel

Volume I · Part a

Heinz Rutishauser

Description of ALGOL 60

Springer-Verlag New York Inc. 1967

Prof. Dr. H. Rutishauser

Eidgenössische Technische Hochschule Zürich

Geschäftsführende Herausgeber:

Prof. Dr. B. Eckmann

Eidgenössische Technische Hochschule Zürich

Prof. Dr. B. L. van der Waerden

Mathematisches Institut der Universität Zürich

ISBN 978-3-642-86936-5   ISBN 978-3-642-86934-1 (eBook)

DOI 10.1007/978-3-642-86934-1

© by Springer-Verlag Berlin · Heidelberg 1967

Softcover reprint of the hardcover 1st edition 1967

Library of Congress Catalog Card Number 67-13537

Titel-Nr. 5118

Preface

Automatic computing has undergone drastic changes since the pioneering days of the early Fifties, one of the most obvious being that today the majority of computer programs are no longer written in machine code but in some programming language like FORTRAN or ALGOL. However, as desirable as the time-saving achieved in this way may be, still a high proportion of the preparatory work must be attributed to activities such as error estimates, stability investigations and the like, and for these no programming aid whatsoever can be of help. In this respect, ALGOL, as an internationally standardized notation which avoids computer-oriented concepts, provides another advantage, not often mentioned, but one which was already the guiding principle at the very beginning of the programming language venture: indeed, a correct ALGOL program is the *abstractum* of a computing process for which the necessary analyses have already been performed. It is the very purpose of this Handbook to establish such abstract formulations of certain computing processes. Therefore, numerical methods given in this Handbook in the form of ALGOL procedures may be put to immediate use wherever ALGOL is known and understood; in fact, application of such a method reduces to little more than calling the corresponding procedure. This, however, requires that ALGOL programs be so designed that they are really abstract in the sense that they do not make use of special properties of a specific computer, and yet take into consideration the general characteristics of digital computation which are (among others): finite precision, finite storage, sequential arrangement of data on an external medium, and — not to be forgotten — finite speed.

Proper use of the procedures published in this Handbook requires of course a thorough knowledge of the language ALGOL which is therefore described in this introductory volume. This description is not given in the style of the ALGOL reports (the most recent ones are reproduced in appendix B of this volume) but was modeled after lectures on ALGOL given at the Swiss Federal Institute of Technology, Zurich. In this way we hope to serve both the beginners as well as the more experienced numerical analyst.

For reasons to be explained later in §4 of this volume, this Handbook sticks to SUBSET ALGOL 60, which is the official IFIP subset of ALGOL; in fact, all programs to be collected in this Handbook shall either be written in this SUBSET or else the deviations shall be clearly stated. Consequently the present volume describes SUBSET ALGOL 60 rather than full ALGOL. In addition to other advantages, this restriction allows us to give a *quantity-oriented* description of the language, which from the standpoint of a prospective user is to be preferred over a *name-oriented* description (which is the only possible way of describing *full* ALGOL).

It is understood that parallel to and following the development of ALGOL, further progress in the field of programming languages (and formal languages in general) has been made. We mention how ALGOL has stimulated the construction of other algorithmic languages as well as the development of new and more efficient methods to translate them into machine code. We may furthermore mention another by-product of ALGOL, namely the introduction of methods for defining algorithmic languages concisely (e.g. the *Backus notation*). However, a comprehensive report on all these activities, as desirable as it might be, would be far beyond the scope of this Handbook, which is intended more as a tool for users of ALGOL than for those interested in programming languages as such. Instead, the reader is referred to Volume I b (A. A. GRAU, U. HILL, and H. LANGMAACK: Translation of ALGOL 60. Edited by K. SAMELSON), which deals with the problem of translating ALGOL text into machine code; at the end of that volume the reader will find an extensive list of references to papers on ALGOL and related topics.

The author is deeply indebted to Mr. F. T. PARKEL for his help in preparing the manuscript and for his invaluable suggestions for improving the text, and to Mr. F. VANNOTTI for testing the ALGOL programs contained in this volume. Furthermore, the author wishes to thank Prof. Dr. E. STIEFEL of the Swiss Federal Institute of Technology for making available the use of the facilities of the computing center.

Zurich, September 1966   H. RUTISHAUSER

Contents

Chapter III

Expressions

Chapter IV

Statements

Chapter V

Miscellaneous Applications

Chapter VI
Declarations

Chapter VIII
Input and Output

Appendix A

Appendix B. The IFIP-Reports on ALGOL

Chapter I

Introduction

§ 1. The Concept of Automatic Programming

"Computing" means to derive, from certain given data, certain results according to given rules. If a computation is done with a desk calculator, one has the rules in mind and applies them as the computation proceeds, taking the intermediate results into consideration. But if the computation must be performed by an automatic computer, all actions to be taken must be planned beforehand and described for the computer as a sequence of instructions, and only then may the computation begin. The entire sequence of instructions is called the *machine program* for that computation, whereas the term *programming* pertains to the preparation of the program.

In the early days of automatic computing, programming was considered as some kind of art, since, indeed, special skill was required to describe the entire computation in advance in a rather queer notation. With the advent of faster computers, however, the need for writing a program for every problem very soon became a nightmare and left no room for artistic feelings. The situation required immediate action in order to reduce the terrible burden. The relief came through the computers themselves: if computers were able to carry such a heavy load of computing, which before had taken years on a desk calculator, they certainly could also assist in writing programs.

Indeed they could; it turned out that it was possible to write programs in a notation somewhere "between" machine code and standard mathematical notation, which was then translated into correct machine code by the computer itself with the aid of a special translation program (usually called *compiler*). Any practice which thus relieves the programmer from writing in machine code is called *automatic programming*. Started around 1950, mainly by M. V. WILKES in England and G. HOPPER in the USA, automatic programming has meanwhile grown into an important branch of computing science, and today a multitude of automatic programming systems are in use which differ widely in scope and efficiency.

The automatic programming systems intended for numerical problems (these are the only ones we are considering here) can be classified into three fairly well separated levels:

a) External machine code

If the instructions which appear inside the computer as strings of digits can be written outside with a mnemonic operation symbol and a decimal address, we call this an external machine code. This is a rather trivial level of automatic programming since the correspondence between the elementary commands in internal and external machine code is practically one to one. Accordingly, the translation process is also extremely simple and sometimes even done by hardware.

b) Assembly languages

Notations of this class are similar to machine code insofar as they also use single, sequentially executed commands of a specified format. However, the addresses of operands and destinations of jumps can be denoted by algebraic symbols which may be suffixed (this for referring to components of vectors). Such methods of automatic programming simplify programming considerably, but on the other hand, the translation into machine code is already fairly complicated since it requires among other things the allocation of addresses.

c) Algorithmic languages

This class contains automatic programming systems which use standard mathematical notation for describing arithmetic operations; in addition it includes certain dynamic elements for describing the flow of a computation. For such languages the programming reduces essentially to writing down the formulae that govern the computation. Since this must be done anyhow, no further reduction can be expected with respect to arithmetics, but great variations in elegance, power and usefulness are still possible within this class by virtue of nonarithmetic features. Quite naturally the translation of an algorithmic language into machine code is extremely complicated, and, accordingly, the corresponding compilers require large storage capacities.

In order to exemplify the three classes of automatic programming systems, we describe here the same piece of computation in the internal and external machine code of ERMETH, in the assembly language CODAP for the CDC 1604A, and in the algorithmic language ALGOL. It should be clear, however, that the merits of the various systems could be brought to light only by examples including loops and subroutine calls.

ALGOL	Internal machine code of ERMETH [1,2]	
$a[101]:= (a[0]+j)\uparrow 2;$	01 1 0000	02 0 0004
$a[1]:= sqrt(1+a[101]);$	19 0 8980	04 0 8980
if $c[-2]<0$ **then goto** $label[150];$	19 1 0101	02 0 9001
goto $label[75];$	21 0 9900	00 0 0000
	19 1 0001	01 3 9998
	22 9 0075	21 9 0150

External machine code of ERMETH [1,2]		CODAP [3]	
A	1,0	LDA	A
+	4	FAD	J
S	8980	STA	TEMP
×	8980	FMU	TEMP
S	1,101	STA	A + 101
+	9001	FAD	ONE
C	9900	STA	TEMP
S	1,1	ENA	TEMP
A	3,9998	RTJ	SQRTF
C+	9,75	RTJ	ERROR
C	9,150	STA	A + 1
		LDA	C − 2
		AJP	M LABEL + 150
		AJP	P LABEL + 75

[1] The electronic computer ERMETH was constructed 1953—1956 under the direction of Prof. E. STIEFEL and Prof. A. P. SPEISER in the Department of Applied Mathematics (Swiss Federal Institute of Technology, Zürich). The ERMETH was in operation from 1956 until 1963; since 1960 it had an ALGOL compiler designed by Dr. H. R. SCHWARZ.

[2] The external and internal operation symbols of the ERMETH are: A|01: clear and add; S|19: store; +|02: floating add; ×|04: floating multiply; C|21: jump; C+|22: jump if positive. The digits following the operation symbol denote index (B-line) and address (these two are separated by a comma in the external notation). At 9900 begins the square root routine (with automatic return), and 9001 contains the floating point constant 1. It is assumed that the addresses of $a[0]$, $c[0]$ and $label[0]$ are stored in index registers 1, 3, 9 respectively, and that j is stored in storage position 4. For more details on ERMETH see J. R. STOCK [36].

[3] The operation symbols of CODAP [11] are: LDA: clear and add; FAD: floating add; FMU: floating multiply; STA: store; RTJ: jump with automatic return; ENA: enter address of operand for the subsequent function call; AJP: conditional jump (P if positive, M if negative).

1*

§ 2. Historical Remarks on Algorithmic Languages

The very first attempt to devise an algorithmic language was undertaken in 1948 by K. Zuse [45]. His notation was quite general, but the proposal never attained the consideration it deserved.

In 1951 the present author tried to show that in principle a general purpose computer could translate an algorithmic language into machine code[1]. However, the algorithmic language proposed in this paper was quite restricted; it allowed only evaluation of simple formulae and automatic loop control (it contained essentially the for-statement of Algol 60). Besides that, the translation method was intermixed with the idea of a *stretched program*, which at that time certainly had some merit as a time-saving device (see [27]) but was not essential for the purpose to be achieved. For these and other reasons this paper did not receive much attention either.

In 1954[2] Corrado Boehm [10] published a method to translate algebraic formulae into computer notation. He considered neither subscripted variables nor loop control, but his method to break up formulae into machine instructions was at this stage a noteworthy step towards the *pushdown methods* described by Samelson and Bauer in [32, 33]. Further early attempts to translate mathematical formulae into machine code were made in 1952 by A. E. Glennie [15] in England and in 1953 by A. A. Liapunov[3] in Russia.

Thus by 1954 the idea of using the computer for assisting the programmer had been seriously considered in Europe, but apparently none of these early algorithmic languages was ever put to actual use.

The situation was quite different in the USA, where an assembly language epoch preceded the introduction of algorithmic languages. To some extent this may have diverted attention and energy from the latter, but on the other hand it helped to make automatic programming popular in the USA. Thus, when in 1954 Laning und Zierler [1] presented their algorithmic language — the first one ever actually used — and shortly thereafter the IBM FORTRAN System [19] was announced, the scientific world was prepared for this new concept.

Meanwhile at Darmstadt an international symposium on automatic computing was held in Oct., 1955[4], where, among other things, algorithmic languages and their translation into machine code were also discussed. Several speakers stressed the need for focusing attention on

---

[1] Lecture at the GAMM (Gesellschaft für angewandte Mathematik und Mechanik) meeting, Freiburg i. Br. March 28—31, 1951. Published in [26].

[2] The paper [10] was officially presented at the ETH on July 10, 1952, as a thesis.

[3] Cited in the introduction to Ershov [13].

[4] The proceedings of this meeting are collected in [18].

unification, that is, on *one universal, machine-independent algorithmic language* to be used by all, rather than to devise several such languages in competition. This became the guiding idea of a working group called the *GAMM Subcommittee for Programming Languages*, which was set up after the Darmstadt meeting in order to design such a universal algorithmic language.

This subcommittee had nearly completed its detailed work in the autumn of 1957, when its members, aware of the many algorithmic languages already in existence, concluded that, rather than present still another such language, they should make an effort towards worldwide unification. Consequently, they suggested to Prof. J. W. CARR, then president of the ACM (Association for Computing Machinery), that a joint conference of representatives of the ACM and the GAMM be held in order to fix upon a common algorithmic language. This proposal received vivid interest by the ACM. Indeed, at a conference attended by representatives of the USE, SHARE and DUO organisations and of the ACM, the conferees had likewise felt that a universal algorithmic language would be very desirable. As a result of this conference, the ACM formed a committee which also worked out a proposal for such a language.

§ 3. The ALGOL Conferences of 1958, 1960, 1962

At that point, direct contact between the GAMM subcommittee and the ACM committee was established through F. L. BAUER in April, 1958, when he presented the GAMM proposal at a Philadelphia meeting of the ACM group. A comparison of the proposals of the ACM and the GAMM indicated many common features. The ACM proposal was based on experience with several successful algorithmic languages. On the other hand, the GAMM subcommittee had worked for a much longer time at their proposal and had from the very beginning the universality of the language in mind.

3.1. ALGOL 58

Both the GAMM and ACM representatives felt that, because of the similarities of their proposals, there was an excellent opportunity for arriving at a unified language. They felt that a joint working session would be very profitable and accordingly arranged for a conference to be attended by four members of the ACM committee and four members of the GAMM subcommittee.

The meeting was held at Zurich, Switzerland, from May 27 until June 2, 1958, and was attended by F. L. BAUER, H. BOTTENBRUCH, H. RUTISHAUSER and K. SAMELSON of the GAMM subcommittee and by

J. BACKUS, C. KATZ, A. J. PERLIS and J. H. WEGSTEIN of the ACM committee[1]. It was agreed that the contents of the two proposals should form the agenda of the meeting and the following objectives were agreed upon:

a) The new language should be as close as possible to standard mathematical notation and be readable with little further explanation.

b) It should be possible to use it for the description of numerical processes in publications.

c) The new language should be readily translatable into machine code by the machine itself.

At this conference it was soon felt that the discrepancies between the notations used in publications on the one hand and the characters available on input/output mechanisms for computers on the other hand were a serious hindrance and might virtually prevent agreement upon a universal algorithmic language. It was therefore decided to disregard printing usage and properties of input/output mechanisms and to focus attention upon an abstract representation (in the sense of a defining standard), called a *reference language,* from which appropriate *publication and hardware* languages might be derived later as isomorphic descendants of the reference language[2]. The notion was, therefore, that reference, publication and hardware languages should be three levels of one and the same language; the conference, however, would then discuss only the reference language. Accordingly, the algorithmic language ALGOL, which was agreed upon at this conference and published in the ALGOL report [5], is defined only on the reference level.

After publication of the ALGOL report [5] much interest in the language ALGOL developed. At the initiative of P. NAUR an ALGOL Bulletin [2] was issued which served as a forum for discussing properties of the language and for propagating its use. The Communications of the ACM introduced an algorithm-section, in which numerical processes are described in terms of ALGOL. Elsewhere ALGOL was also used more and more for describing computing processes.

---

[1] In addition to the members of the conference, the following persons participated in the preparatory work of the committees: GAMM: P. GRAEFF, P. LÄUCHLI, M. PAUL, F. PENZLIN; ACM: D. ARDEN, J. McCARTHY, R. RICH, R. GOODMAN, W. TURANSKI, S. ROSEN, P. DESILETS, S. GORN, H. HUSKEY, A. ORDEN, D. C. EVANS.

[2] For the relation between reference, publication and hardware language see [5]. It should be recognized that experience has shown that ALGOL programs may well be published in the reference language, and therefore extra publication languages are in fact unnecessary. On the other hand, hardware languages have proved necessary to such an extent that it was decided to standardize a few carefully selected hardware representations of the ALGOL symbols (cf. section 7.4).

On the other hand it was soon found that certain definitions given in the ALGOL-58 report were either incomplete or even contradictory or otherwise unsatisfactory for the description of numerical processes. As a consequence many proposals were made to remove these defects[1].

3.2. ALGOL 60

In view of the constructive criticism that evolved and the proposals made, it was decided that another international ALGOL conference should take place. Accordingly, the GAMM subcommittee organized a preliminary meeting at Paris in Nov. of 1959, attended by about 50 participants from Western Europe, from which 7 delegates for the final ALGOL conference were selected. The ACM committee likewise selected 7 delegates at a preparatory meeting held in Washington D.C. at the same time. Both the European and the USA delegation made proposals for removing the inconsistencies from the ALGOL report and also for making changes in the language. These proposals took the criticisms as much as possible into consideration.

The conference, held at Paris, Jan. 11—16, 1960, was attended by J. W. BACKUS, F. L. BAUER, J. GREEN, C. KATZ, J. McCARTHY, P. NAUR, A. J. PERLIS, H. RUTISHAUSER, K. SAMELSON, B. VAUQUOIS, J. H. WEGSTEIN, A. V. WIJNGAARDEN, M. WOODGER[2]. The proposals worked out prior to the conference again formed the agenda of the meeting, but in addition the conferees had at their disposal a completely new draft report prepared by P. NAUR, which served as a basis for discussion during the conference.

From the beginning it was obvious that rather than just adding a few corrections to ALGOL 58, it was necessary to redesign the language from the bottom up. This was done, and accordingly ALGOL 60, as the language emerging from the Paris conference is officially called, was in many respects entirely different from ALGOL 58. It is defined on the reference level in the ALGOL-60 report [6] edited by P. NAUR. Since publication of this report, ALGOL 58 has become obsolete, but many of its features have been carried over into other algorithmic languages.

3.3. The Rome amendments of 1962

Soon after the ALGOL-60 conference a number of inconsistencies were again found in the new ALGOL report. Most of them were just mistakes, but others led to discussions which revealed a considerable conceptual

---

[1] Most of these proposals have been published in the Comm. of the ACM, Vol. 2 (1959) and/or in the ALGOL Bulletin Nr. 7.

[2] W. TURANSKI of the American delegation was fatally injured in an automobile accident just prior to the Paris conference.

divergence among the experts, and it proved impossible to bridge the gap between the opposite interpretations of the ALGOL-60 report. It was therefore decided to seize the opportunity when most of the members of the ALGOL-60 conference would be attending an IFIP (International Federation for Information Processing) meeting at Rome to discuss these discrepancies. A formal ALGOL meeting, attended by F. L. BAUER, J. GREEN, C. KATZ, P. NAUR, K. SAMELSON, J. H. WEGSTEIN, A. V. WIJN-GAARDEN, M. WOODGER, R. KOGON, R. FRANCIOTTI, P. Z. INGERMAN, P. LANDIN, M. PAUL, G. SEEGMUELLER, R. E. UTMAN and W. L. V. D. POEL was held on April 2—3, 1962, at Rome.

At this meeting the known mistakes and inconsistencies were corrected as far as agreement could be obtained, and a corrected report, called the *Revised* ALGOL *Report* [7] (in the following abbreviated RAR), was issued under the auspices of the IFIP, which meanwhile had taken over the responsibility for the further maintenance and development of ALGOL. At the same time a list of the corrections was published [40].

However, even the Revised ALGOL Report leaves the following questions still open:

a) Side effects of function designators.

b) The *call by name*-concept.

c) Static or dynamic own-concept?

d) Static or dynamic for-statement?

e) Conflicts between specifications and declarations.

§ 4. ALGOL Dialects and the IFIP Subset of ALGOL 60

Partly because of the uncertainty caused by the still unsettled issues of ALGOL as mentioned in 3.3 above, partly because some of the more sophisticated features of ALGOL 60 are hard to implement, only few (if any) compiler makers have so far implemented (i.e. built compilers for) the full language ALGOL 60. The worst effect of this is not that the language cannot be fully used — in fact ALGOL offers still enough advantages to make its implementation highly desirable — but that different implementors make different restrictions and thus create many dialects of ALGOL.

Usually such dialects exclude the use of the more sophisticated features of ALGOL (recursive procedures, the own-device) besides restricting the use of types (cf. 8.1). But only too often such dialects have been "enriched" by non-ALGOL features such as the *format declaration of FORTRAN* and the use of lists. As useful as such extensions may seem for the individual user, the net effect is that they destroy the universality of the language, and this is much worse than certain inconveniences in the use of the present ALGOL. To correct this very unsatisfactory situation, it was concluded that, insofar as many implementors are virtually forced to make restrictions, they should at least be urged to make identical restrictions and of course no extensions.

To achieve this, it was decided that the IFIP should issue an official set of restrictions which define a true subset of ALGOL 60 as defined by the RAR. This subset should be easier to implement and not contain the controversial concepts but still be sufficient for most numerical applications.

At a first meeting of the ALGOL *working group* of IFIP (W. G. 2.1), held on August 28-30, 1962, in Munich, the then available subsets SMALGOL [4] and ALCOR [9] were reviewed and the possibilities for coalescing them were discussed. At further meetings of the W. G. 2.1 at Delft (Sept. 10—13, 1963) and Tutzing (March 16—20, 1964) the final decision was made as to what features of ALGOL 60 should be excluded from the official IFIP subset. Since then the IFIP council has approved this decision and released a *Subset Report* [20] (in the following abbreviated as SR) which defines the official IFIP subset through restrictions added to the full language as defined by the RAR. The official name of this subset is SUBSET ALGOL 60.

This official subset excludes the controversial features of ALGOL 60 but is still sufficient for most numerical applications. It is contained in most other subsets that have been created hitherto; accordingly, most existing compilers can translate programs written in SUBSET ALGOL 60. This is one of the reasons why this Handbook sticks to the subset: Despite the nearly babylonic confusion with respect to the abilities and restrictions of the presently existing ALGOL compilers, the programs published in this Handbook will run successfully with most existing ALGOL compilers without further adjustments. This, as we know, is far from being true for ALGOL programs that make extensive use of the more sophisticated features of ALGOL 60.

But there are also other reasons why we restrict ourselves to SUBSET ALGOL 60: If one uses only this subset, one may design a special subset-compiler, which, because it need not take care of the controversial concepts of ALGOL, will require less storage space and produce more efficient object programs.

§ 5. Preliminary Definition of ALGOL

The description of a complete calculation in terms of ALGOL is called an ALGOL *program*. It consists of a sequence of statements and declarations which are separated from each other by semicolons, the whole being enclosed by **begin** and **end**. Upon execution of the ALGOL program, the statements call for certain well-defined actions, whereas declarations serve merely to state the occurrence of certain quantities.

In this section we describe some of the features of ALGOL, however without being strict or complete.

5.1. Arithmetic expressions and assignment statements

5.1.1. The elimination of the quadratic term from a cubic equation $x^3 + a x^2 + b x + c = 0$ is achieved by a substitution $x = y - \dfrac{a}{3}$ which leads to the equation $y^3 + p y + q = 0$, where the new coefficients p, q are given as $p = b - a^2/3$, $q = c - ab/3 + 2a^3/27$. In ALGOL the computation of p, q for given values of a, b, c is described as

$$p := b - a \uparrow 2/3;$$

$$q := c - a \times b/3 + 2 \times a \uparrow 3/27;$$

This is not a complete ALGOL program but just a small section of one, namely two *assignment statements*. They are executed in the order given and thus produce the coefficients of the reduced equation. The action to be taken by the corresponding object program in the computer is: Take the values a, b, c from the respective storage positions, compute p and q and store the resulting values in the positions reserved for p and q.

5.1.2. The general idea of such an assignment statement is that the value of the variable on the left side of the *assignment symbol* := is computed as defined by the formula on the right side. Such formulae, in ALGOL called *arithmetic expressions*, are written essentially in standard mathematical notation (except that the multiplication symbol $\times$ may not be omitted and that an exponentiation symbol $\uparrow$ is used instead of raising the exponent) and use the following elements:

a) *Operation symbols* $+ - \times / \uparrow$ and *parentheses* ( ).

b) *Numerical constants*, e.g.

   131   .0325   34.5678 $135_{10} - 12$ $9.87654_{10}3$,

with a decimal point (not a comma!) as separator between integral and fractional part, and (if needed) with a scaling factor expressed as a power of ten (the symbol $_{10}$ indicates the beginning of the exponent part).

c) *Simple variables* represented by arbitrary names like

  x p ys $k1$ *apex*   *zr105*   *j777j*   *alpha*   *certif*,

which are strings of letters, possibly including decimal digits, but always beginning with a letter. Such a variable represents the value that has been assigned to it before.

d) Subscripted variables. In contrast to the usual subscript notation, e.g. a_i, b_{k+l}, $c_{2,j}$, d_{j_k}, subscripts are written in ALGOL on the normal line level and enclosed in *brackets*. Accordingly the above examples — which denote components of vectors, matrices, etc. — appear in ALGOL as

$$a[i] \qquad b[k+l] \qquad c[2,j] \qquad d[j[k]],$$

and $v[k] := 0$ is the ALGOL equivalent of setting the k-th component of a vector $\vec{v}$ to zero.

e) Standard functions. The 9 names

$$sin \quad cos \quad sqrt \quad ln \quad exp \quad arctan \quad abs \quad sign \quad entier$$

refer to the so-called standard functions. As an example $sin(x+y)$ represents the sine of $x+y$, and $entier(x/2)$ means the largest integer not exceeding $x/2$.

5.1.3. Let us now give a few examples:

a) The length of a vector (x_1, x_2, x_3) in 3-dimensional space is given by the formula $s = \sqrt{x_1^2 + x_2^2 + x_3^2}$. In ALGOL s is computed by the assignment statement

$$s := sqrt(x[1] \uparrow 2 + x[2] \uparrow 2 + x[3] \uparrow 2)$$

b) The angle *alpha* of a triangle given with sides a, b, c is defined as

$$tan\left(\frac{alpha}{2}\right) = \sqrt{\frac{(s-c)(s-b)}{s(s-a)}}, \quad \text{where } 2s = a+b+c.$$

In ALGOL we have to compute and assign s first (because only then can its value be used) before evaluating the expression defined by the given equation for *alpha*:

$$s := (a+b+c)/2;$$
$$alpha := 114.5915\,5903 \times arctan(sqrt((s-c) \times (s-b)/(s \times (s-a))));$$

(*alpha* in hexagesimal degrees but with decimal fractions).

Thus it appears that ALGOL is somewhat less compact and elegant than standard mathematical notation, but this is the price to be paid in order that all symbols can be written straightforward on the normal line level (which of course is an indispensable requirement for mechanical reading devices).

c) Statements like

$$k := k+1$$

are allowed in ALGOL. This statement increases the value of k by 1. Indeed, it says that the value of k is taken, 1 added to it, and the result is again assigned to k. Of course, the previous value of k is destroyed by this operation.

5.2. For-statements

If all components of a vector $\vec{v}$ must be set to zero, the operation $v[k] := 0$ must be performed once for every one of the values $k=1$, 2, ..., n. In ALGOL we have a shorthand notation for this:

$$\textbf{for } k := 1 \textbf{ step } 1 \textbf{ until } n \textbf{ do } v[k] := 0$$

In fact, as the prefix **for** ... **do** says, it performs the statement $v[k] := 0$ n times with the prescribed values of k.

As a further example, consider the calculation of the length of a vector $\vec{v}$ in n-space. If n is variable, the expression $v_1^2 + v_2^2 + \cdots + v_n^2$ cannot be transcribed directly into ALGOL but must be evaluated by a summation loop: Let $s = v_1^2 + v_2^2 + \cdots + v_{k-1}^2$, then the next partial sum is obtained by adding v_k^2 to s, and this can be described in ALGOL as $s := s + v[k] \uparrow 2$. To obtain the sum, we must simply repeat this operation for $k = 1, 2, \ldots, n$, and start with $s = 0$:

$$s := 0 \ ;$$
$$\textbf{for } k := 1 \textbf{ step } 1 \textbf{ until } n \textbf{ do } s := s + v[k] \uparrow 2 \ ;$$

The general rule is that a for-statement consists of a *for-clause* (the part between **for** and **do** inclusive) followed by a statement. The for-clause says how often and for what values of the *controlled variable* (this is the variable following **for**) the statement should be executed.

It should be recognized, however, that the controlled variable need not be a subscript (although it usually is) but can also take on non-integer values. As an example, the following two statements compute the sum of the values of the function e^x/x at $x = 0.01, 0.02, \ldots, 10.00$:

$$s := 0;$$
$$\textbf{for } x := 10 \textbf{ step } -0.01 \textbf{ until } 0.01 \textbf{ do } s := s + exp(x)/x$$

Here we let x run backwards in order to reduce the influence of the roundoff errors in the summation. But since now the controlled variable is of real type, also its stepping causes rounding errors, which may lead to incorrect termination of the loop (cf. 30.5.1).

5.3. Compound statements

If a more complicated computation, i.e. a sequence of statements rather than a single statement, must be executed repeatedly, these statements must be enclosed by **begin** and **end**[1], and the for-clause must

---

[1] Since the semicolons are not constituents of the single statements but rather separators between them, no semicolon is required before the **end**.

be placed in front of the **begin**, e.g.

 for $k := 1$ **step** 1 **until** $n - 1$ **do**
 begin
  $e[k] := (e[k]/q[k]) \times q[k+1];$
  $q[k+1] := (q[k+1] - e[k]) + e[k+1]$
 end

In fact, by embracing an arbitrary sequence of statements (these may be all sorts of statements) by **begin** and **end**, a new kind of statement, called *compound statement*, is created. Therefore, if such a compound statement is preceded by a for-clause, it is by definition repeated as a whole. Thus the above example has for $n = 4$ the same effect as the following sequence of statements:

$$e[1] := (e[1] / q[1]) \times q[2];$$
$$q[2] := (q[2] - e[1]) + e[2];$$
$$e[2] := (e[2] / q[2]) \times q[3];$$
$$q[3] := (q[3] - e[2]) + e[3];$$
$$e[3] := (e[3] / q[3]) \times q[4];$$
$$q[4] := (q[4] - e[3]) + e[4];$$

Likewise the multiplication of an $n \times n$-matrix $A = (a[i, k])$ with a vector $\vec{b} = (b[i])$, yielding a vector $\vec{c} = (c[i])$, is described by

 for $i := 1$ **step** 1 **until** n **do**
 begin
  $s := 0;$
  for $k := 1$ **step** 1 **until** n **do** $s := s + a[i, k] \times b[k];$
  $c[i] := s$
 end

Indeed, every execution of the compound statement computes one, namely the i-th, component of the product vector.

5.4. Labels and goto-statements

Normally the statements of an ALGOL program are executed in the order in which they are written. However, this order may be interrupted by placing a goto-statement, e.g.

goto *maior*

at the place where the interruption should take place and a corresponding *label* and *colon*,

maior:

in front of the statement with which the computation should proceed after the interruption.

Consider, for instance, the sequence of statements

$$entry: \quad aux := (a+b)/2;$$
$$b := sqrt(a \times b);$$
$$a := aux;$$
$$\textbf{goto } entry$$

Assume now that execution of the program in which these statements are embedded has proceeded to the label *entry*. Then the next three statements are executed, after which **goto** entry is encountered. This has the effect that the computation begins again after the label *entry*, from which it proceeds again downwards. Thus obviously those four statements are executed over and over again with infinite repetition; i.e. the computation is caught in a *closed loop*, from which it can be freed only by intervention of the operator.

Such closed loops can be avoided by making jumps conditional, as will be shown below in 5.5 and later in Chapter IV (in fact, goto-statements are only useful in conjunction with conditional statements).

5.5. The if-statement

The execution of a statement can be made conditional by placing an *if-clause* in front of it, e.g.

$$\textbf{if } x = 0 \textbf{ then } x := {}_{10} - 20$$

This obviously means that $x := {}_{10} - 20$ is executed if and only if x was exactly zero and may serve to avoid trouble in a later division. Likewise

$$\textbf{if } abs(a-b) > {}_{10} - 9 \textbf{ then goto } entry$$

makes the jump to *entry* subject to the condition $|a-b| > {}_{10} - 9$ and thus could be used to prevent the closed loop in the example given in 5.4 above.

If an if-clause is placed in front of a compound statement, then the execution of the whole is subject to the condition stated in the if-clause. As an example,

```
if abs(t) < ln(1 + theta) then
begin
    y := 0;
    for p := m step −1 until 0 do y := y × t + c[p]
end
```

evaluates the polynomial $c_0 + c_1 t + c_2 t^2 + \cdots + c_m t^m$ (and assigns the result to y) if and only if $|t| < \ln(1 + theta)$.

In the general case, an if-clause may be placed in front of any statement *which is not already an if-statement*. In other words, the construction **if** $x = 0$ **then if** $z > \sin(y)$ **then** ... is forbidden. However, it is also unneeded since ALGOL permits achieving the intended effect by *conjunction* of the two conditions:

$$\textbf{if } x = 0 \wedge z > \sin(y) \textbf{ then} \ldots$$

This if-clause means that the statement following **then** is executed if and only if *both* conditions $x = 0$ and $z > \sin(y)$ hold (compare, however, 28.4.3).

5.6. If-else-statements

Sometimes one wants to extend the if-statement by saying also what should be done if the condition stated in the if-clause does *not hold*. To achieve this, we do not end the if-statement with a semicolon or **end** but with a symbol **else**, and then append another statement, e.g.

$$\textbf{if } p > q \textbf{ then } f := p \textbf{ else } f := q$$

In such a case, always one of the two statements is executed, namely the first if the condition holds, the second if the condition does not hold. Thus the above statement assigns the larger of the two values p, q to the variable f.

In the general case, the if-else-statement may have the form

$$\textbf{if } C \textbf{ then } U \textbf{ else } S$$

where C denotes a condition, U any statement not beginning with **if** or **for**, and S any statement whatsoever.

An application of this possibility is shown by the following piece of program which describes the computation of the largest real root of a cubic equation $x^3 + a x^2 + b x + c = 0$. It should be recognized, however, that this is still not a complete program since it contains neither declarations nor input- and output-operations:

```
begin
    p := b − a ↑ 2/3;
    q := c − a × b/3 + 2 × a ↑ 3/27;
    disc := (q/2) ↑ 2 + (p/3) ↑ 3;
    if q = 0 ∧ p > 0 then x := − a/3
else
    if q = 0 ∧ p ≦ 0 then x := − a/3 + sqrt(− p)
```

first alternative of an if-else-statement

second alternative

else
    if $disc > 0$ **then**
    begin
        $u := sqrt\,(disc) + abs\,(q/2)\,;$ third
        $v := u \uparrow 0.3333\,3333\,33333\,;$ alternative
        $x := -a/3 - (v - p/(3 \times v)) \times sign\,(q)$
    end
else
    begin
        $phi := arctan\,(2 \times sqrt\,(-disc)/q)\,;$
        if $q > 0$ **then** $phi := phi - 3.1415\,9265\,3589\,79\,;$ fourth
        $x := -a/3 + 2 \times sqrt\,(abs\,(p/3)) \times cos\,(phi/3)$ alternative
    end
end

It should be recognized that in a case like $p = 1$, $q = 0$, where the conditions for both the first and third alternative hold, by definition only the first alternative is actually executed.

5.7. Declarations

A full ALGOL program is essentially a sequence of statements which are separated from each other by semicolons, the whole being enclosed by **begin** and **end**. In addition, however, the sequence must contain *declarations* between the first **begin** and the first statement of the program; these serve to state certain properties of the variables (and other quantities) occurring in the program. Example:

begin
    real $x, y, t, pz, gyrosc\,;$
    integer $i, j, k\,;$
    array $a\,[1:9], b, c, d\,[0:5], u, v\,[1:20, 1:30]\,;$ declarations
    integer array $f, g\,[-1:6]\,;$
    $x := -1.5\,;$ statements of
    $\vdots$ the program
end

In this program, the declarations at the beginning state that the real (floating point) variables x, y, t, pz and $gyrosc$ and the integer-valued (fixed point) variables i, j, k will occur. In addition, the arrays a, b, c, d, f, g with one subscript (vectors) and the arrays u, v with two subscripts (matrices) are declared and can therefore be used in the program. The numbers in the brackets denote the lower and upper bounds for the subscripts of these arrays, e.g. the first subscript of the arrays u, v can run from 1 to 20, the second from 1 to 30.

These declarations do not necessarily mean that all the declared variables actually occur in the program or that all components of the declared arrays must actually be used. But a variable which has not been declared cannot be used, and a component of an array whose subscript is beyond the declared limits cannot be used.

To be precise, declarations can appear not only after the first **begin** of a program, but also after a later **begin** (this introduces the block-concept, for which see § 42). In an array declaration following a later **begin** the subscript bounds may depend on calculated values, e.g.

begin
   integer n, m, p;
   $n := entier \, (sqrt \, (1 + exp \, (6)))$;
   begin
      array $a[1:n, 1:n]$;
      :
      :

5.8. Complete programs

An ALGOL program cannot be complete unless it also contains statements which perform the transfer of initial data and results of a computation from and to the outside world. In ALGOL such *input-* and *output-*operations may be performed via the *standard I/O-procedures* (for the details of which see Chapter VIII). For the moment let us see how the cubic-equation program of 5.6 may be completed by adding the necessary declarations and I/O-operations:

begin
   real $a, b, c, p, q, u, v, phi, disc, x$;
   $inreal \, (1, a)$;   } input of the coefficients a, b, c
   $inreal \, (1, b)$;   via channel 1.
   $inreal \, (1, c)$;

                       } Insert here the piece of
                       program given in 5.6

   $outreal \, (2, x)$ output of x via channel 2
end

Chapter II

Basic Concepts

The definition of a programming language consists of several parts, namely

a) The definition of the *basic symbols*, which are the atoms of the language.

b) The *syntax* (or syntactic rules); these are the rules which define how the basic symbols can be concatenated to larger units (in the following called *syntactic objects*) and finally to complete ALGOL programs.

c) The *semantics* (or semantic rules), i.e. the rules which define what actions a given ALGOL program (or section hereof) should initiate at execution time.

In the following a semiformal definition of the language SUBSET ALGOL 60[1] on the reference level is given together with examples which show the properties and possibilities of the language. Since the subsequent text deals exclusively with SUBSET ALGOL 60, the word ALGOL will from now on automatically pertain to this subset, while the term *full* ALGOL will be used where, as an exception, reference to the language defined by the RAR must be made.

§ 6. Auxiliary Conventions

6.1. Syntactic forms

A new class Γ of syntactic objects will be introduced by its *syntactic form*, which is a sequence of basic symbols and/or capital letters[2]. The syntactic form defines the general element of a new class of syntactic objects as follows: Basic symbols represent themselves, but a capital letter stands for an arbitrary element of the respective class of syntactic objects. Where the new class is defined by several syntactic forms $\Sigma_1, \Sigma_2, \ldots$, it is the union of all subclasses defined by each of the Σ_i.

Take for instance the syntactic form $I[E]$, where I denotes the class "identifier" and E the class "arithmetic expression". Obviously this defines a class of syntactic objects which consist of an identifier followed

---

[1] The official definition of this subset is given only by the two reports RAR and SR together.

[2] According to the SR (section 2.1) the capital letters are not basic symbols of SUBSET ALGOL 60 and are therefore available for this purpose.

by an arithmetic expression in brackets (subscripted variables with one subscript), e.g.

$$delo\,[k], \qquad merit\,[x+1], \qquad ccit\,[(k+1)\times k/2], \qquad d\,[d\,[5]].$$

6.2. French quotes

In order to distinguish between ordinary text and ALGOL text, ALGOL programs and parts thereof (down to basic symbols) as well as syntactic forms will from now on be enclosed by French quotes « »:

$$\texttt{«goto } entry\texttt{»}, \qquad \texttt{«}p\texttt{»}, \qquad \texttt{«}aux := (a+b)/2\texttt{»}, \qquad \texttt{«}I[E]\texttt{»}.$$

It should be clear, however, that this is by no means a rule of ALGOL but only an ad hoc construction for avoiding confusion.

6.3. The ellipsis

The ellipsis ... will be used in syntactic forms in the sense of *obvious continuation* in order to indicate that certain parts of the syntactic form occur with an unspecified number of repetitions, including the degenerate case where the number of repetitions is only one. Accordingly a construction such as

$$\texttt{«}I[E, E, \ldots, E]\texttt{»}$$

represents the union of all syntactic forms

$$\texttt{«}I[E]\texttt{»}, \qquad \texttt{«}I[E, E]\texttt{»}, \qquad \texttt{«}I[E, E, E]\texttt{»}, \qquad \texttt{«}I[E, E, E, E]\texttt{»}, \quad \text{etc.}$$

6.4. The syntactic diagram[1]

Besides the means described above, new syntactic objects will also be defined more precisely by *syntactic diagrams*. The class of objects to be defined is designated by its name in a bold frame, and the arbitrary element of this class is obtained by running from the origin *o* in an arbitrary way along the arrows to the bold frame, whereby the basic symbols found in the circles and rounded boxes and arbitrary elements of the classes listed in the rectangular boxes are collected and aligned in the order in which they are met. As an example

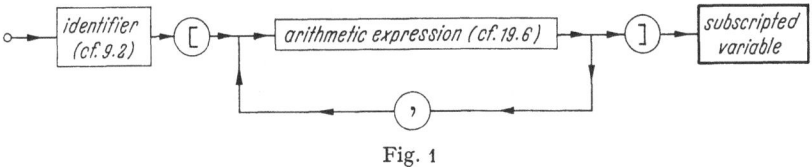

Fig. 1

---

[1] Dipl. Ing. A. SCHAI, Director of the Computing Center of the ETH, Zurich, proposed this hitherto unpublished modification of the Burroughs Syntactical Chart (cf. Comm. ACM, Sept. 1961, pp. 393).

defines a new class of syntactic objects which consist of an identifier followed by an arbitrary number of arithmetic expressions which are separated by commas and enclosed in brackets. It defines therefore the same class as the syntactic form «$I[E, E, \ldots, E]$», namely the class "subscripted variable".

6.5. Undefined situations

The semantic rules sometimes state that the outcome of a certain operation or the effect of executing a certain piece of ALGOL program is *undefined*. This simply means that whenever such a situation is encountered in a computation, the further execution of the program is unpredictable and in fact may produce any effect a computer is capable of. Such a piece of program is therefore incorrect.

On the other hand, it is not necessarily an error if an ALGOL program produces an *undefined value* during its execution, provided this value is not further used by the program.

§ 7. The Basic Symbols of ALGOL

7.1. Set of basic symbols

The set of basic symbols of ALGOL contains[1]:

a) All small *letters* of the Roman alphabet:

«a», «b», «c», «d», «e», «f», «g», «h», «i», «j», «k», «l», «m», «n», «o», «p», «q», «r», «s», «t», «u», «v», «w», «x», «y», «z».

b) The *decimal digits* «0», «1», «2», «3», «4», «5», «6», «7», «8», «9».

c) The *logical constants* «**true**» and «**false**».

d) The *arithmetic operators* «$+$», «$-$», «$\times$», «$/$», «$\uparrow$».

e) The *relational operators* «$=$», «$\neq$», «$<$», «$>$», «$\leq$», «$\geq$».

f) The *logical operators* «$\neg$», «$\wedge$», «$\vee$», «$\supset$», «$\equiv$».

g) The *sequential operators* «**goto**», «**if**», «**then**», «**else**», «**for**», «**do**».

h) The *separators* «,», «.», «$_{10}$», «:», «;», «:=», «⊔», «**step**», «**until**», «**while**», «**comment**».

i) The *brackets* «(», «)», «[», «]», «'», «'», «**begin**», «**end**».

j) The *declarators* «**real**», «**integer**», «**Boolean**», «**array**», «**switch**», «**procedure**».

k) The *specificators* «**label**», «**string**», «**value**».

---

[1] According to the SR, the symbols $\div$, **own** and all capital letters are not basic symbols of the subset.

7.2. Delimiters

The elements of groups d) through k) are usually called *delimiters*, the others are non-delimiters. In addition, some of the basic symbols have individual names:

/	*slash* or *solidus*		:=	*assignment symbol*
,	*comma*		⊔	*space symbol* (this is
.	*period*			used only in strings).
$_{10}$	*base ten* [1]		( )	*parentheses*
:	*colon*		[ ]	*brackets*
;	*semicolon*		' '	*string quotes*.

The 23 underlined[2] English words among the basic symbols are called *word-symbols*. They have been incorporated into the language because it was felt that the readability of ALGOL programs would be improved if for certain nonarithmetic operations such word-symbols expressing the action to be performed were chosen instead of unusual symbols like ⋈ ⚈, etc. It should be recognized, however, that underlining expresses the fact that the word-symbols are *atoms* of the language like other basic symbols and therefore can be neither decomposed nor translated into other national languages.

7.3. Typography

Where an ALGOL program is written on paper, it is understood that the order of the basic symbols in the ALGOL text is the same as the conventional order of letters in the plain English text. However, blank space[3], change to a new line and indenting of the latter have no significance in ALGOL. These devices are syntactically nonexistent in ALGOL and can therefore be used freely to improve the readability of ALGOL programs without changing their effect. Extensive use of this possibility has been made in this volume.

7.4. Hardware representations

Few I/O-devices for electronic computers accept all basic symbols of ALGOL as given in 7.1 above. Most users of ALGOL are therefore forced to take recourse to so-called hardware representations for entering ALGOL programs into the computer. This means that they must replace the nonavailable symbols occurring in ALGOL text with suitable com-

---

[1] In order to distinguish it from the number « 10 », the *base ten* should be written below the line level.

[2] Because of the difficulty of achieving underlining in printed text, bold face (grotesque type) is used throughout this volume instead of underlining.

[3] It should be recognized that the space symbol «⊔» used in strings is considered different from a unit of blank space; indeed the former has a very definite meaning.

binations of other symbols, but of course only such combinations can be used which cannot occur otherwise in an ALGOL program. Examples:

:=   It is general practice to represent this symbol by a colon, followed by an equality symbol.

[For I/O by punched cards this symbol is usually represented by the combination (/.

≦   In the Bull Gamma 60 computer represented by <=.

The word-symbols are also a problem, since underlining is usually not possible with I/O-mechanisms. To resolve this difficulty, a specific symbol not contained in the set of basic symbols (e.g. the apostrophe or $) is chosen as an *escape symbol* with the convention that any word enclosed between a pair of escape symbols is considered as underlined. Thus «**begin**», «**procedure**» may be represented by 'begin', $ procedure $.

The escape symbol is also used for representing other non-available symbols, e.g. 'less' as representative of « < ».

The following hardware representations (for punched cards and 5-channel paper tape) have been accepted as a DIN *standard* [12]. The table given indicates for every basic symbol of SUBSET ALGOL 60 either the punching combination for cards and tapes, or else the character combination used to circumscribe the basic symbol. For punched cards the conventional enumeration of the punched card rows, i.e. 12–11–0–1–2–3–4–5–6–7–8–9 applies, while for paper tape the 5 channels are enumerated 1–2–3–4–5, with the sprocket hole between channels 2 and 3. Note that paper tape has the peculiarity that the same punching combination may denote two different symbols, depending upon whether the mechanism is on letter-shift (BU) or figure-shift (ZI). Two extra punch combinations are reserved for changing the shift.

a) Symbols which can be punched directly.

Symbol	Tape (*ZI*)	Punched cards	Symbol	Tape (*ZI*)	Punched cards
0	2–3–5	0	—	1–2	11
1	1–2–3–5	1	×	1–4–5	11–4–8
2	1–2–5	2	/	1–3–4–5	0–1
3	1	3	,	3–4	0–3–8
4	2–4	4	(	1–2–3–4	0–4–8
5	5	5	)	2–5	12–4–8
6	1–3–5	6	10	3–5	
7	1–2–3	7	:	2–3–4	
8	2–3	8	;	1–2–4	
9	4–5	9	[	1–3–4	
=	2–3–4–5[1]	3–8[1]	]	2–4–5	
.	3–4–5	12–3–8	'	1–3[2]	4–8[2]
+	1–5	12			

[1] = is used only as constituent of the assignment symbol «:= ».
[2] ' (apostrophe), used as *escape symbol*.

Symbol	Tape (BU)	Punched cards	Symbol	Tape (BU)	Punched cards
a	1–2	12–1	n	3–4	11–5
b	1–4–5	12–2	o	4–5	11–6
c	2–3–4	12–3	p	2–3–5	11–7
d	1–4	12–4	q	1–2–3–5	11–8
e	1	12–5	r	2–4	11–9
f	1–3–4	12–6	s	1–3	0–2
g	2–4–5	12–7	t	5	0–3
h	3–5	12–8	u	1–2–3	0–4
i	2–3	12–9	v	2–3–4–5	0–5
j	1–2–4	11–1	w	1–2–5	0–6
k	1–2–3–4	11–2	x	1–3–4–5	0–7
l	2–5	11–3	y	1–3–5	0–8
m	3–4–5	11–4	z	1–5	0–9

b) Auxiliary symbols for 5-channel tape.

Symbol	Punch comb.	Meaning
WR	4	carriage return
ZL	2	line shift
ZWR	3	space
ZI	1–2–4–5	figure shift
BU	1–2–3–4–5	letter shift

c) Symbols represented by combinations of other symbols:

Symbol	Representation on tape	Representation on punched cards	Symbol	Representation on tape and punched cards
10		'	↑	'power'
:		..	<	'less'
;		.,	>	'greater'
[		(/	$\leqq$	'not greater'
]		/)	$\geqq$	'not less'
:=	:=	.=	=	'equal'
‹	'('	'('	$\neq$	'not equal'
›	')'	')'	$\neg$	'not'
⊔	ZWR	(blank)	∧	'and'
			∨	'or'
			⊃	'impl'
			≡	'equiv'

d) The 23 word-symbols are all represented by enclosing the words in apostrophes instead of underlining them, i.e. «**begin**» is represented by 'begin', «**procedure**» by 'procedure', etc.

§ 8. Values

Upon execution of an ALGOL program, certain well-defined actions take place, the most frequent one being that operations are performed upon certain values, whereby other values are produced. These are again involved in operations, etc., until the «**end**» of the program is reached. All other operations serve solely to assist in these calculations.

8.1. Types of values

The values upon which an ALGOL-program can operate fall into three classes[1]:

a) The values of type **real**[2], i.e. the class of real values.

b) The values of type **integer**, i.e. the class of all integers.

c) The values of type **Boolean** (the logical values **true** and **false**).

Concerning the integers, it should be recognized that the same integer value can either be of integer or real type, depending on how it was generated. Which of the two cases actually occurs in a given situation, is defined by *rules of type* at appropriate places in the following chapters. The distinction between real and integer type is important, since in critical cases the results of a computation may depend on it.

8.2. Computer limitations

In actual computing, values of real and integer type must be represented by digital numbers (usually real type values by *floating point* numbers, integer type values by *fixed point* numbers). Consequently, such values are subject to computer limitations, i.e. they can be represented only if they remain between certain bounds, and values of type **real** can be represented at best only approximately. Since these are facts that we could not hope to alter, it was indispensable that they somehow be built into the framework of the language ALGOL, otherwise we would have been being utterly unrealistic. As a consequence the following rules have been accepted:

a) A value of type **real** is considered as inherently inaccurate, i.e. as being defined only with finite precision, and arithmetic operations performed with them must be assumed to be affected by (usually small) roundoff-errors.

b) Values of both types **real** and **integer** can be represented only if they remain within certain (computer-dependent) bounds, and it must

---

[1] Despite the wording used in the RAR, labels are not values in this sense. Accordingly, we use here — without changing the content — a different wording which does not give labels the status of values.

[2] It has become customary to say *the value x is of type T* instead of *the value x belongs to class T*.

be taken into account that the further course of the calculation is un-defined as soon as a value exceeds its respective bounds (so-called *over-flow*).

c) On the other hand, it is assumed that arithmetic operations per-formed with values of type **integer** are carried out exactly, provided the result is again of type **integer** and remains within the prescribed bounds.

8.3. Consequences of computer limitations

According to what has been said above, it must be tolerated that an ALGOL program may produce results which deviate from the expected theoretical values, or that its execution may be discontinued because of overflow[1]. Even worse, the same ALGOL program will usually produce different results with different computers, or cause overflow on com-puter A but not on computer B.

This seems a hopeless situation, and ALGOL does not give the slightest hope for overcoming these difficulties. On the contrary, it is entirely up to the numerical analyst to design an ALGOL program such that it pro-duces useful results on any computer despite the above-mentioned short-comings. But how this should be achieved is not a question of ALGOL and therefore is not treated here, except that we shall indicate in some of the programming examples what can be done to overcome the dif-ficulties associated with computer limitations (cf. § 36).

§ 9. Quantities and their Names

Whenever a programmer describes a computation in terms of ALGOL, he automatically introduces certain quantities which are abstract objects distinguished by their names. They serve to facilitate the description of the program but obtain their meaning through the program itself (in actual computation, these quantities are realized as storage positions or groups of storage positions).

9.1. Kinds of quantities

The following quantities are used in ALGOL:

9.1.1. A *simple variable* is an object to which a value may be assigned and then remains associated with it until a further assignment to the same variable.

---

[1] It should be recognized that some of the programs published in this Handbook presume that in case of *underflow* (i.e. the exponent of a floating point number exceeds its lower bound) at worst the machine representation of "floating zero" is produced. Several of these programs will not work properly with computers that produce arbitrary effects upon underflow.

9.1.2. An *array* is a set of elements, called the *components* of the array, every one of which behaves like a simple variable. The components of an array are distinguished by a set of p integers (subscripts) $i_1, i_2, \ldots, i_p$, where p is called the *dimension of the array*. If we interpret the subscripts as coordinates in a p-dimensional space, then the entire array corresponds to the total of all unit-gridpoints in a p-dimensional hyperbox

$$l_k \leq i_k \leq u_k \qquad (k = 1, 2, \ldots, p),$$

whose boundaries (i.e. the *array bounds* $l_1, l_2, \ldots, l_p, u_1, u_2, \ldots, u_p$) are given in the corresponding array declaration (cf. § 39).

9.1.3. A *label* is a designation given to a specific spot in an ALGOL program.

9.1.4. A *switch* is a one-to-one correspondence between an ordered set of n labels and the integers $1, 2, \ldots, n$.

9.1.5. A *procedure* is an operator which can operate upon other quantities (e.g. compute certain results from given arguments). However, the properties of procedures can differ markedly from the properties of mathematical functions and operators; in fact, procedures more often resemble the subroutines in ordinary machine-code programming.

9.2. Identifiers

Quantities can appear in ALGOL programs only through their names, which are syntactic objects classified as *identifiers*.

9.2.1. Examples of identifiers are

$$\text{«}x7\text{»}, \quad \text{«}a\text{»}, \quad \text{«}y\text{»}, \quad \text{«}phi\text{»}, \quad \text{«}vcrit\text{»}, \quad \text{«}pt77tp\text{»}.$$

9.2.2. Identifiers have the syntactic form

$$\text{«}X\text{»} \quad \text{or} \quad \text{«}XYY \ldots Y\text{»},$$

where X stands for an arbitrary letter and every Y means an arbitrary letter or digit. Thus an identifier is a sequence of letters and/or digits, but always beginning with a letter. Syntactic diagram:

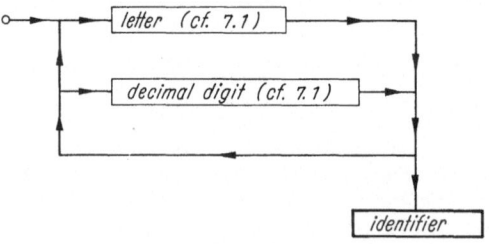

Fig. 2

The possibility of forming names of more than one symbol gives a sufficient supply of names. It allows to circumscribe the capital and greek letters (*bigm*, *beta*), and to give variables mnemonic names such as *vcrit*. On the other hand it excludes the implied multiplication, since e.g. «*a b*» is always considered as one name and never as the product of *a* with *b*.

9.2.3. *Semantics*. Identifiers may be chosen freely and have, with the exception of the identifiers of the standard functions and standard I/O-procedures[1], no preassigned meaning. However, the same identifier cannot be used to denote more than one quantity at once (for more details see § 42, Semantics of Blocks).

9.2.4. *Restriction*. Identifiers may be of arbitrary length, but only the leading six characters of an identifier are used for identification (see SR, item 2.4.3). Thus two identifiers which agree in the first six characters, e.g.

«*output17*» and «*outputvalue*»,

are considered identical in the subset (but not in full ALGOL!) and therefore may cause trouble if both are used in the same ALGOL program. In order to avoid trouble of this sort, it is strongly recommended to restrict the length of identifiers whenever possible to at most six characters.

9.3. Scope of a quantity

With the exception of labels and the reserved identifiers, every quantity used in an ALGOL program must be declared. Such a declaration, besides announcing the quantity and the name used for it, defines also other properties (for this see Chapter VI) and especially the *scope* of a quantity. The latter is defined as that part of an ALGOL program in which the quantity exists and can be called through its identifier. Outside the scope the quantity is either nonexistent or temporarily inaccessible.

§ 10. Numerical Constants

Values appear in ALGOL programs usually as values of variables; these values can be changed in the course of a calculation. However, where a value is known a priori and is the same in all applications of the program, it can be given directly as a *numerical* or *logical constant*.

The syntactic objects denoting numerical constants are the *unsigned numbers*, with the important subclasses *unsigned integers*, *decimal numbers*, *exponent parts*. The logical constants are represented by the basic symbols «**true**» and «**false**».

---

[1] Lists of these reserved names are given in 15.2.1 for the standard functions and in 49.1 for the standard I/O-procedures.

10.1. Examples of unsigned numbers

Unsigned integers: «0», «175», «1», «3014», «000».
Decimal numbers: «.197», «510.0», «0000.0070», «11.754».
Exponent parts: «$_{10}85$», «$_{10}-7$», «$_{10}0$», «$_{10}4711$».
General case: «$175_{10}-7$», «$0007_{10}003$», «$1.1_{10}11$», «$00.00_{10}+00$»,
$$«1.234567_{10}89».$$
(Some of these examples are inflated by insignificant zeroes, but this is allowed.)

10.2. Syntax

10.2.1. The *unsigned integers* have the syntactic form

$$«ZZ \ldots Z»,$$

where every Z represents an arbitrary decimal digit.

10.2.2. With this, the *unsigned numbers* have one of the following syntactic forms (the G's denote arbitrary unsigned integers):
Decimal numbers: «G», «$.G$», «$G.G$»[1].
Exponent parts: «$_{10}G$», «$_{10}+G$», «$_{10}-G$».
General case: a decimal number followed by an exponent part[2].

10.2.3. *Syntactic diagram* (see Fig. 3).

10.3. Semantics

An unsigned number is a syntactic object which always represents the same numerical value. Decimal numbers have the conventional meaning, whereas the exponent part is a scaling factor expressed as a power of ten.

ALGOL imposes restrictions neither upon the length of numerical constants nor upon the size of the numbers represented by them, but of course the computer limitations mentioned in 8.2 apply.

10.4. Types

Unsigned integers represent values of type **integer**, while all other numerical constants are of type **real**. As a consequence, «$2_{10}2$» and «200.000» are of type **real**, while «200» is of type **integer** but represents the same value. The logical constants «**true**» and «**false**» are of type **Boolean**.

---

[1] The comma, which in some European countries is used as the standard separator between integer and fractional part of a number, cannot be used for that purpose in ALGOL. On the other hand, it is also forbidden to insert commas as digit group separators in long numbers, e.g. 1,234,567.89.

[2] Besides this, the RAR mentions — without making further use of it — a syntactic entity called *number*.

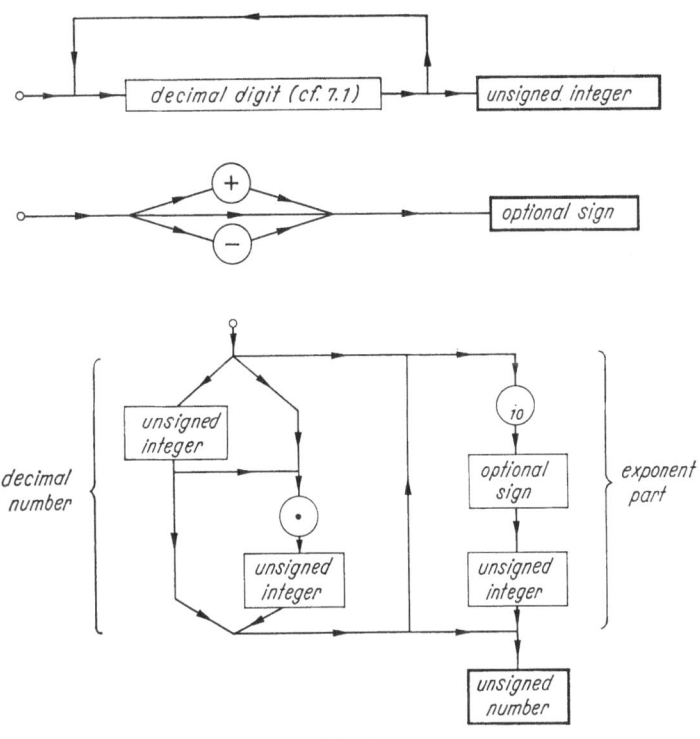

Fig. 3

10.5. Negative constants

Unsigned numbers always represent non-negative values. Where negative numerical constants are required, a symbol «−» may be placed in front of an unsigned number; it should be recognized, however, that such combinations, e.g. «$-_{10}6$», «$+1.234567_{10}89$», are no longer unsigned numbers, but *arithmetic expressions,* for which different rules apply (cf. §16). On the other hand, a sign following the symbol «$_{10}$» is a constituent of the exponent part and not an arithmetic operator.

§ 11. Labels

Labels are used in ALGOL programs for identifying the destinations of jumps, e.g. we can write «**goto** *arica*» and place a corresponding label and colon «*arica* :» in front of the statement to which the jump should be directed. Sometimes labels are also used to mark statements just for explanatory purposes.

11.1. Syntax

A label is an arbitrary identifier[1] and therefore a sequence of letters and/or digits, but always beginning with a letter. Accordingly the following are examples of labels:

$$\ll arica\gg, \quad \ll x7\gg, \quad \ll a\gg, \quad \ll dixi\gg, \quad \ll pt77tp\gg.$$

11.2. Source- and destination labels

Depending on the synactic position in which it appears in an ALGOL program, a label is either

a *destination label*, i.e. a label in front of a statement, e.g.

$$\ll arica : x := 1\gg, \text{ or}$$

a *source label*,   if it occurs in a goto-statement, e.g. «**goto** *x7*», in a switch list, or as actual parameter.

In the following, the word *label* will usually refer to the label as an entity without regard to its syntactic position, whereas the attributives *source* and *destination* indicate a specific syntactic position of the label in question.

11.3. Semantics

A *destination label* is itself a quantity; it marks a spot in an ALGOL program. A *source label* on the other hand is not itself a quantity but only referring to the corresponding destination label.

As a rule, a source label can refer to a destination label only if the two match exactly symbol for symbol (they are then called *corresponding*). However, the restriction of length (cf. 9.2.4) holds also for labels, and thus a jump «**goto** *identical*» can well have the labelled statement «*identity* : $z := x + sin(y)$» as its destination.

11.4. Scopes of labels

Since destination labels are quantities, the rules of scope must be observed. As an example, the same identifier may not be used more than once as destination label at the same block level; it may, however, occur several times as source label. For further details see § 42.

§ 12. Strings

In order that arbitrary sequences of basic symbols can be handled by an ALGOL program (mainly for controlling I/O-operations), strings

---

[1] According to the SR, item 3.5.1, unsigned integers are not admitted as labels in SUBSET ALGOL 60.

have been introduced. However, strings can appear in ALGOL programs only as actual parameters of procedure statements or function designators (cf. §15, § 26).

12.1. Examples

$$\texttt{«'}..this \sqcup is \sqcup a \sqcup \text{'} string \text{''} \texttt{»},$$
$$\texttt{«'} \sqcup\sqcup\sqcup\sqcup\sqcup \texttt{'»},$$
$$\texttt{«'} \sqcup //57 \dots - ([[) \sqcup \textbf{'else'} \sqcup a := b \sqcup\sqcup \texttt{'»},$$
$$\texttt{«'} \sqcup\sqcup s \sqcup ddd.ddd_{10} + dd \sqcup\sqcup \texttt{'»}.$$

12.2. Syntax

A string has the syntactic form

$$\texttt{«' '»} \quad \text{or} \quad \texttt{«'} QQ \dots Q \texttt{'»},$$

where «'» and «'» are the string quotes and every Q denotes either itself a string or any basic symbol except the string quotes.

Syntactic diagram:

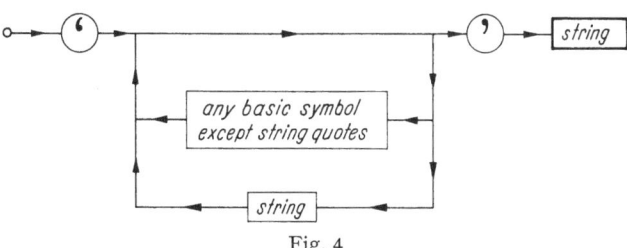

Fig. 4

Note. Except for the rules given above, the symbols contained in a string are completely arbitrary; indeed, all other syntactic rules do not apply inside strings.

12.3. Semantics

Strings serve as actual parameters of procedure statements or function designators[1] to whose execution they contribute certain nonarithmetic information, e.g. formats for printing. Otherwise the basic symbols contained in strings have no bearing on the execution of an ALGOL program.

Within strings spaces may have a meaning. In order to discriminate between relevant and irrelevant spaces, the former are denoted by the symbol «⊔». This symbol, however, cannot be used outside strings.

---

[1] For the precise conditions under which a string may appear as actual parameter of a procedure call see 45.3.1.

§ 13. Comments

In order that explanations may be given between statements and declarations of an ALGOL program, the following rule has been adopted:

13.1. The comment convention

After any symbol «;» or «**begin**» occurring in an ALGOL program, arbitrary text may be inserted, provided the latter is enclosed between the separators «**comment**» and «;». Indeed, the symbols following the symbol «**comment**» up to and including the next following semicolon are considered as nonexistent.

Likewise the symbols following «**end**» up to, but excluding, the next following «;», «**end**» or «**else**» (whichever comes first) are considered as nonexistent.

It should be recognized, however, that «**comment**» may never be placed after a symbol other than «;» or «**begin**».

13.2. Examples

«**begin comment**: the variable *zeta* is no longer used ;»,
«; **comment** time is $2 \times money$. stop forever ;»,
«**end** of loop ;»,
«**end** of type $27c$-branching **end**»,
«**end** of while-condition $// + / + //$ — — **else**».

According to the above conventions, these 5 pieces of program are equivalent to the constructions

   «**begin**»,   «;»,   «**end** ;»,   «**end end**»,   «**end else**»

respectively.

13.3. Conflicting situations

If the ALGOL report is taken literally, the construction

   «**end begin comment** *jan* ; *klaus* ;»[1]

gives rise to an ambiguity. Indeed, depending on whether we consider first the comment situation induced by «**comment**» or by «**end**», the above example will be equivalent to

   «**end** ;»   or to   «**end** ; *klaus* ;».

However, if we accept that ALGOL programs are always read strictly from left to right, then the «**end**» is considered first, hence «**begin**»

---

[1] See DIJKSTRA, E. W.: ALGOL-Bulletin [2] Nr. 12, item 12.1.

and «**comment**» are nonexistent and therefore this example must be interpreted as «**end** ; *klaus* ;». Similarly, other examples such as

«**comment** *look at* ' ⊔; $x := 1$⊔' *prime* $:= 0$;»,
«*prtext*(' **comment** ⊔ *nonsense*') ;»,
«**comment begin comment** a ; b ;»

obtain an unambiguous meaning by reading strictly from left to right. All the same it seems advisable not to make use of conventions which are not explicitly stated in the RAR. It is therefore recommended that the symbols «**comment**», «**end**», «'», and «'» not be used in a comment situation, i.e. in one of the situations described above.

Chapter III

Expressions

An expression describes the computation of a new value from other, already given values, in an obvious notation. The given values appear in the expression

a) directly as numerical or logical constants;
b) as values of variables;
c) as values of function designators.

Expressions fall into two classes:

Arithmetic expressions compute values of real or integer type; they are the backbone of all numerical computations.

Boolean expressions compute values of type **Boolean**. Basically their purpose is to facilitate logical calculations (Boolean algebra), but they are also used frequently in numerical calculations when decisions concerning the further course of the calculation must be made.

Wherever an expression is encountered during execution of an ALGOL program, it is evaluated according to the rules given in this chapter and produces a single value (of either real or integer or Boolean type). How this value is further used is not defined by the expression but by the context, i.e. the statement or declaration into which the expression is embedded.

Though the rules adopted in ALGOL for evaluation of an expression are mainly in accordance with long established conventions, they will be defined here again from the bottom up. These definitions are necessarily recursive because the rules for subscripted variables and function designators already use the concept of an expression. Furthermore, function designators cannot be fully defined before procedures have been introduced, but those in turn use all other elements of the language.

§ 14. Variables

Variables serve as carriers of values. Indeed, a computed value can be attached to a variable and then remains associated with it. This value can be used later in the calculation simply by inserting the variable wherever the value is required.

We distinguish *simple variables*, which are naked identifiers, and *subscripted variables*, which are identifiers appended by subscripts; the latter, however, are not written below the line as usual but are enclosed in brackets.

Every variable has a certain type (**real**, **integer**, **Boolean**) which defines the type of value that can be attached to it.

14.1. Examples

14.1.1. «*omikro*», «*delta*», «*t*», «*x1*», «*c878z*» are simple variables.

14.1.2. «$r[k]$», «$r[k+1]$», «$r[5]$», «$r[m \times n - k\uparrow 2]$» are subscripted variables representing the k-th, $k+1$-th, fifth and $(mn-k^2)$-th component of the array r.

14.1.3. «$q[ji, r[k], -3]$», «$rax[rax[rax[1]]]$», «$i[j[m, n]]$» are subscripted variables with nested subscripts. If e.g. in the first of the examples above we have $ji=8$, $k=2$, $r[2]=5$, then it represents the $(8, 5, -3)$-component of the three-dimensional array q. Furthermore, if the components of the array rax have the values $rax[j]=2j$, then the second example represents the value 8.

14.2. Syntax

14.2.1. *Simple variables* have the syntactic form «I», i.e. they are just identifiers.

14.2.2. *Subscripted variables* have the syntactic form

$$«I[E, E, \dots, E]»,$$

where I represents an arbitrary identifier (more precisely, I represents the name of the array of which the subscripted variable is a component), and the E's are *subscript expressions*, i.e. ordinary arithmetic expressions[1], but with a special rule of evaluation (for which see § 20).

14.2.3. Both kinds of variables are described by the following syntactic diagram:

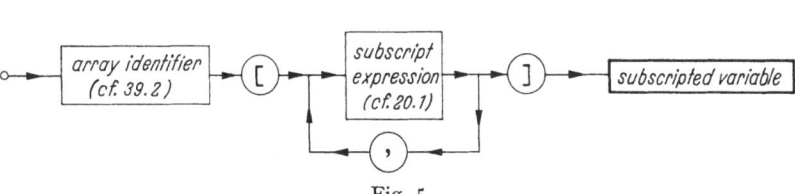

Fig. 5

---

[1] Note that according to § 16, E can also mean a simple variable or numerical constant because these are special cases of arithmetic expressions.

14.2.4. Both simple and subscripted variables can appear in ALGOL programs in essentially the same syntactic positions, except that a subscripted variable can appear neither as actual parameter, if the corresponding formal parameter is called by name (SR, item 4.7.3.2), nor as controlled variable in a for-statement (SR, item 4.6.1).

14.3. Semantics

14.3.1. A *simple variable*, if encountered during execution of an expression[1], represents a single value, namely the value that has most recently been assigned to the variable. This value is time-dependent insofar as it is changed by every new assignment to the variable; between consecutive assignments, however, the value of a variable remains constant.

14.3.2. A *subscripted variable* «$I[E_1, E_2, \ldots, E_p]$», if encountered in an expression[1], represents also a single value defined as follows: Evaluate the subscript expressions $E_1, E_2, \ldots, E_p$; if their values are $i_1, i_2, \ldots, i_p$, then the subscripted variable represents the value that has most recently been assigned to the $i_1, i_2, \ldots, i_p$-component of the array I.

14.3.3. *Restrictions.* The use of variables is subject to certain restrictions which have to do with the fact that simple variables and arrays must be declared and have *scopes* (cf. 42.2). In fact a simple variable is nonexistent and unusable outside its scope. Furthermore, a subscripted variable is nonexistent not only outside the scope of the corresponding array, but also if any one of the subscript expressions produces a value outside the bounds prescribed by the corresponding array declaration for that subscript position. Finally, the value of a simple or subscripted variable is undefined before the first assignment to it has occurred.

14.4. Types

Every variable is of a certain type (**real, integer, Boolean**), which simply means that the variable can only represent values of that type. However, in contrast to FORTRAN, types of variables are not distinguished syntactically but are defined by corresponding declarations. More precisely, the type of a simple variable is defined by a *type declaration* (cf. § 38), while for subscripted variables the type is defined for the whole array (hence is common to all its components) by an *array declaration* (cf. § 39).

§ 15. Function Designators

A function designator is a syntactic object which initiates evaluation of a certain function. The resulting function value is then used in the

---

[1] For the meaning of a simple or subscripted variable appearing on the left side of an assignment statement see § 21; for their use as actual parameters of procedure calls cf. § 45, 46.

evaluation of the expression in which the function designator occurred. Two classes of functions are used in ALGOL:

a) The 10 *standard functions*

   sin, cos, exp, ln, sqrt, arctan, abs, sign, entier, length,

which are permanent constituents of the language and have a fixed meaning. These 10 functions can be used in ALGOL programs without being declared and their identifiers are *reserved names* which should not be used otherwise.

b) *Function procedures.* Besides the standard functions, the user may introduce any functions he finds useful. However, these must be declared by corresponding *function procedure declarations* (cf. § 46).

15.1. Examples

a) For the standard functions:

   «*arctan* (1)», «*sin* (1.45 + z)», «*ln* (1 + x)»,
   «*length* ('..this ⊔ is ⊔ a ⊔ 'string'')».

b) For function procedures:

   «*pi*», «*radius*»   (these are without argument),
   «*bessel* (n, x)», «*sinhyp* ($_{10}$—3 × theta + exp (x))»,
   «*decide* (15, **true**, roda)», «*decide* (n, a = entier (b), w)»,
   «*bessel* (n↑2, sqrt (v [entier (x)]))»

(The reader may find declarations for some of these functions in § 46).

15.2. Syntax

The syntactic object which causes evaluation of a function is the function designator. It can be used as a *primary* in an arithmetic or Boolean expression (cf. § 16, § 18) and therefore is itself a complete arithmetic or Boolean expression and may appear wherever the syntax allows for an expression.

A function designator has one of the following syntactic forms:

15.2.1. For the standard functions:

$$«I(Q)»,$$

where I denotes a *standard function identifier*, i.e. one of the 10 names

   «*sin*», «*cos*», «*exp*», «*ln*», «*sqrt*»,
   «*arctan*», «*abs*», «*sign*», «*entier*», «*length*»,

and Q represents either an arbitrary arithmetic expression (for the functions *sin* through *entier*), or a string or a string identifier (for the function *length*).

15.2.2. For the function procedures (compare also 41.2):

«I»,   if the corresponding procedure declaration has the
                               syntactic form «T **procedure** I ; S»,

«$I(A_1, A_2, \ldots, A_p)$», if the procedure declaration has the syntactic form
                               «T **procedure** $I(F_1, F_2, \ldots, F_p)$; VCS».

Hereby T represents one of the declarators «**real**», «**integer**», «**Boolean**», while I denotes the *procedure identifier* (the name of the function) and the A's are the *actual parameters,* which may be either expressions or identifiers or strings (see § 45, 46 for further conditions which actual parameters must meet).

15.2.3. *Structurized forms.* It is permitted to replace any of the commas separating the actual parameters of a function designator with a syntactic construction «$)XX \ldots X:($» (parameter delimiter), in which the X's denote arbitrary letters.

15.2.4. *Syntactic diagram:*

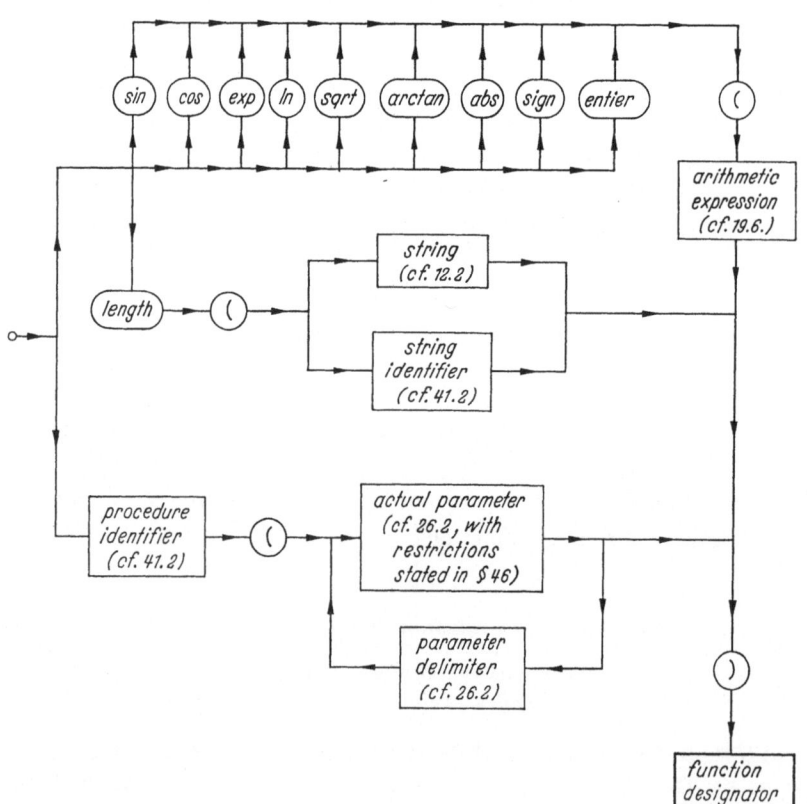

Fig. 6

15.3. Semantics

A function designator F, if encountered during evaluation of an expression E, represents a single value which is obtained by evaluating F intercalatively in the evaluation of E as follows:

15.3.1. For a function designator pertaining to one of the standard functions *sin* through *entier*, the arithmetic expression appearing as its argument is evaluated, and with that value x the function is evaluated as indicated by the following table:

Function	Value of function designator $I(x)$	Type of function value
$sin(x)$	conventional value, x taken in radians	**real**
$cos(x)$	idem	**real**
$exp(x)$	e^x	**real**
$ln(x)$	natural logarithm, undefined if $x \leqq 0$	**real**
$sqrt(x)$	positive branch of $\sqrt{x}$, undefined if $x < 0$	**real**
$arctan(x)$	conventional value y, $-\pi/2 < y < \pi/2$	**real**
$abs(x)$	$|x|$	**real**
$sign(x)$	$+1$ if $x > 0$, -1 if $x < 0$, 0 if $x = 0$	**integer**
$entier(x)$	integer value k such that $k \leqq x < k+1$	**integer**

15.3.2. The value of «*length*(S)» (where S represents a string) is the number of basic symbols contained between the outermost string quotes of the string S, and is of type **integer**.

15.3.3. For the evaluation of function designators corresponding to declared function procedures see § 46.

15.4. Types

Every function is of a certain type, which means that corresponding function designators can produce only values of that type. The types of the standard functions are defined in 15.3.1 and 15.3.2 above: The functions *sign*, *entier* and *length* are of type **integer**, while all other standard functions are of type **real**. The type of a function procedure is defined by the type declarator in front of the corresponding function procedure declaration (cf. 41.2.2).

§ 16. Simple Arithmetic Expressions

Arithmetic expressions serve to describe computations in the domain of real or integer values. They are written in conventional mathematical notation, except that for exponentiation an operation symbol «↑» is

used instead of raising the exponent, and that the multiplication symbol
« × » may not be omitted.

Arithmetic expressions may be *simple* or *conditional*. The latter are
treated in § 19, while here we deal only with simple arithmetic expressions.

16.1. Examples

«*a*»

«*sin* (1.45 + *z*) »

«*r* [*k*] »

«1.234567$_{10}$ − 89»

«9↑9»

«*a* + *b*»

«1/*sqrt* (*y*[1]↑2 + *y*[2]↑2 + *y*[3]↑2)↑3 »

«2.137$_{10}$ − 5 × *b* + *c*/(*a* + 7 × *b*) − *delta*↑2/4.1 + *a* × *b*/*v*↑2»

«(*a* + *b*) × *c* − *d*/(*e*/*f* + *g*) × *h*↑*k* − *l*/*m*»

«0.25 × *sqrt* ((*a* + *b* + *c*) × (−*a* + *b* + *c*) × (*a* − *b* + *c*) × (*a* + *b* − *c*)) »

Note. According to the syntax, also unsigned numbers, variables and
function designators are themselves simple arithmetic expressions, and
therefore all examples given in 10.1, 14.1, 15.1 are special cases of simple
arithmetic expressions.

16.2. Syntax

16.2.1. A *simple arithmetic expression* has the syntactic form

$$«S \quad P \; O \; P \; O \; P \; ... \; P \; O \; P»,$$

where the *P*'s are the *primaries* of the expression (these may be either
unsigned numbers, simple or subscripted variables, function designators,
or arithmetic expressions in parentheses) and the *O*'s denote arithmetic
operators (cf. 7.1, d). *S* represents one of the symbols «+», «−» or
blank space.

16.2.2. *Terms and factors.* If all additive operators (those for addition
and subtraction) which are not enclosed in parentheses or brackets are
removed from an arithmetic expression, the latter falls into pieces which
we call the *terms* of the expression. Likewise, a term falls into pieces
called *factors* if all multiplicative operators (those for multiplication and
division), as far as they are not inside brackets or parentheses, are removed from it.

As an example, the terms of the eighth example in 16.1 are

«2.137$_{10}$ − 5 × *b*»,   «*c*/(*a* + 7 × *b*) »,   «*delta*↑2/4.1 »,   «*a* × *b*/*v*↑2»,

while «*a*», «*b*», «*v*↑2» are the factors of the last term.

16.2.3. *Syntactic diagram*:

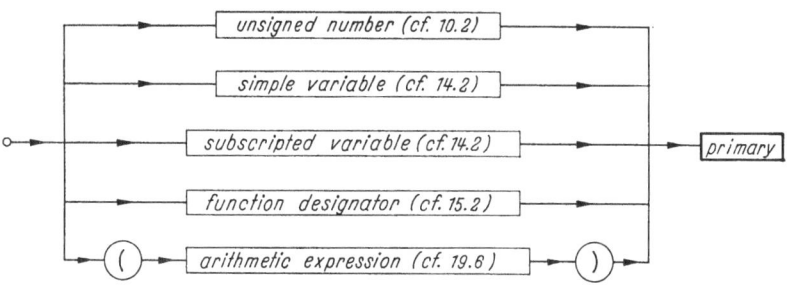

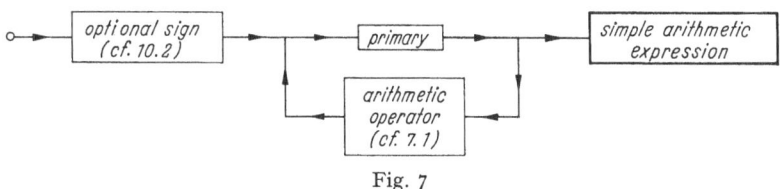

Fig. 7

16.3. Semantics

16.3.1. An arithmetic expression, whenever it is encountered during the execution of an ALGOL program, produces a value according to the following rules:

a) The primaries of the expression are evaluated independently and their values are used for further evaluation.

b) Outside the primaries the following *rules of precedence* apply:

First $\uparrow$
Second $\times$ and $/$
Third $+$ and $-$

c) Except for b), precedence of arithmetic operators goes *from left to right*.

d) Expressions occurring within the primaries are evaluated according to the same rules.

The precise content of these rules may also be expressed by the

16.3.2. *Equivalence rule for arithmetic expressions*: Let

$$\text{«}P_0\ O_1\ P_1\ O_2\ P_2\ \ldots\ P_{n-1}\ O_n\ P_n\text{»}\,[1]$$

be a simple arithmetic expression in which the P's represent the primaries and the O's are the arithmetic operators connecting them. If O_k is an

---

[1] It is assumed that the *optional sign* in front of this expression is blank space, otherwise we would place another primary 0 (zero) in front of the expression and thus establish the required syntactic form without changing the meaning.

operator which in the sense of 16.3.1, b has precedence over O_{k-1} (or if $k=1$), and O_{k+1} does *not* have precedence over O_k (or if $k=n$), then the above expression is defined to be equivalent to

$$\text{«}P_0\ O_1\ \ldots\ O_{k-1}\ (P_{k-1}\ O_k\ P_k)\ O_{k+1}\ \ldots\ O_n\ P_n\text{».}$$

Iterated application of this rule permits parenthesizing an arithmetic expression to such an extent that the order in which the operations should be carried out becomes obvious. In this way the eighth example of 16.1 is transformed into

$$\text{«}(((((2.137_{10}-5\times b)+(c/(a+(7\times b))))-((delta\uparrow 2)/4.1))+((a\times b)/(v\uparrow 2)))\text{»,}$$

which simply means that the original expression could be evaluated in the following order:

$$\text{«}2.137_{10}-5\times b+c/(a+7\times b)-delta\uparrow 2/4.1+a\times b/v\uparrow 2\text{»}$$

$$
\begin{array}{cccccc@{\qquad}cc@{\quad}ccc}
| & | & | & | & | & | & | & | & | & | & | & | \\
1 & 5 & 4 & 3 & 2 & 8 & 6 & 7 & 12 & 9 & 11 & 10
\end{array}
$$

Where function designators and/or subscripted variables are involved in an arithmetic expression, evaluation of the former and selection of the correct array component for the latter is also considered as an operation which must be ranged among the other operators, e.g.:

$$\text{«}-7.394_{10}7\uparrow iso\,[k+2\times j]\times bessel\,(n\uparrow 2,\ sqrt\,(x\uparrow 2+y\uparrow 2))/rax\,[rax\,[1]]\text{».}$$

$$
\begin{array}{ccccccccccccc}
| & & | & | & | & | & | & | & | & | & | & | & | & | \\
15 & & 4 & 3 & 2 & 1 & 14 & 10 & 5 & 9 & 6 & 8 & 7 & 13 & 12 & 11
\end{array}
$$

It should be recognized, however, that what these examples show is not the only possible order of evaluation. Indeed, the compiler maker has, within the bounds prescribed by the equivalence rules, considerable freedom for organizing the evaluation of arithmetic expressions.

16.3.3. *Execution of single arithmetic operations, types*

Let a and b denote two single arithmetic values (of type **real** or **integer**). Then the operations

$$\text{«}a+b\text{»,}\quad \text{«}a-b\text{»}\quad\text{and}\quad \text{«}a\times b\text{»}$$

are defined according to convention. The resulting value is of type **integer** if and only if both operands are of type **integer**, otherwise of type **real**.

The operation «a/b» is the ordinary division of real numbers. The resulting value is of type **real** irrespective of the types of a and b, and therefore one can never rely upon the precision of the quotient of two integers, even if the result is (theoretically) again integer-valued. Besides, the quotient is *undefined* if $b=0$.

The operation «$a \uparrow b$» denotes exponentiation, a being the base and b the exponent. The result c of this operation is defined by the following table:

		b of type real			b of type integer			Legend:
		$>$	$=$	<0	$>$	$=$	<0	C: Conventional value of a^b of type **real**.
a of type **real**	>0	C	E	C	C	E	C	E: 1.0 of type **real**.
	$=0$	O	U	U	O	U	U	O: 0.0 of type **real**.
	<0	U	U	U	C	E	C	U: Undefined.
a of type **integer**	>0	C	E	C	X	X	U	X: Undefined, except if b is an unsigned integer; then $a \uparrow b$ has
	$=0$	O	U	U	X	U	U	the conventional value and is
	<0	U	U	U	X	X	U	of type **integer**.

16.3.4. *Undefined situations.* Besides the cases mentioned already, the value of a simple arithmetic expression is undefined if any one of its primaries has an undefined value or a value of type **Boolean**.

16.4. Type of the value of a simple arithmetic expression

According to the rules given above, the value of a simple arithmetic expression is (if defined at all) of type **integer** if and only if the values of all primaries are of type **integer** and if neither divisions nor exponentiations (other than with unsigned integers as exponents) occur. In all other cases the value is of type **real**.

16.5. Confrontation of examples with conventional notation

Conventional notation	ALGOL notation
a_{p_q}	$a[p[q]]$
$a_{i,k}$	$a[i, k]$
a^{b^c}	$a \uparrow (b \uparrow c)$
a^{n+1}	$a \uparrow (n+1)$
$\dfrac{-b + \sqrt{b^2 - 4ac}}{2a}$	$(-b + sqrt\ (b \uparrow 2 - 4 \times a \times c)) / (2 \times a)$
$\dfrac{1}{2\sqrt{t}} e^{-\frac{x}{4t}}$	$exp\ (-x/(4 \times t))/(2 \times sqrt\ (t))$
$\dfrac{1}{\sqrt{1 - \sin^2\left(\dfrac{\alpha}{2}\right) \sin^2 \psi}}$	$1/sqrt\ (1 - sin\ (alpha/2) \uparrow 2 \times sin\ (psi) \uparrow 2)$

§ 17. Relations

A relation is a predicate which produces a logical value as the result of comparing two arithmetic values. It can appear in a logical expression, and the logical value produced by it is used in the evaluation of this expression (see § 18 below).

17.1. Examples

$«x = 0»$ $\begin{cases}\text{This produces the value } \textbf{true} \text{ if } x \text{ is zero, otherwise} \\ \text{it yields the value } \textbf{false}.\end{cases}$

$«a{\uparrow}2 + b{\uparrow}2 < 1»$ $\begin{cases}\text{Test whether the point } a, b \text{ lies inside the unit} \\ \text{circle; if so, it produces the value } \textbf{true}.\end{cases}$

$«i \neq n + 1»$

$«sin\,(a/f) \leq v\,[entier\,(bessel\,(n, x) \times 10)]»$

$«n = 2 \times entier\,(0.25 + n/2)»$ $\begin{cases}\text{The result of this relation is } \textbf{true} \\ \text{if } n \text{ is even, } \textbf{false} \text{ if } n \text{ is odd.}\end{cases}$

$«-1 > x1crit»$.

17.2. Syntax

A relation consists of two simple arithmetic expressions separated by one of the 6 relational operators (cf. 7.1):

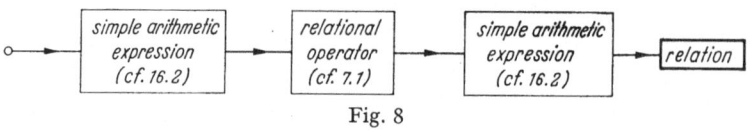

Fig. 8

The simple arithmetic expressions appearing on either side of a relational operator are sometimes called the *comparands* of the relation.

17.3. Semantics

17.3.1. A relation describes a condition between two arithmetic expressions. Wherever encountered during execution of a program, it represents a logical value, namely

the value **true**, if the condition is fulfilled,
the value **false**, if the condition is *not* fulfilled.

To obtain this value, the comparands of the relation are evaluated indepedently and only then are their values compared.

17.3.2. *Influence of roundoff errors.* If a, b are variables of type **real**, then roundoff errors may completely reverse the value of a relation such as $«a = b»$, which is **true** if and only if the current values of a and b coincide *exactly*. Analogous effects occur with the other five relational operators.

All the same, such a relation $«a = b»$ may be useful, e.g. for skipping parts of a computation in which exact coincidence of the numerical values of a and b would be disastrous. However, it should not be overlooked that it may make a difference which of the relations $«a = b»$ or

«$a - b = 0$» is used for this purpose, since due to roundoff errors no one can guarantee that these two relations will produce the same logical value.

The influence of roundoff errors may become a severe problem where termination of iterative computing processes is done by tolerance criteria, because these must be expressed in ALGOL by relations. It is not further tragic if the roundoff errors involved in the evaluation of such a relation have the effect that the termination is only delayed by a few iteration steps, but it might happen that a termination criterion such as «$abs(x) < eps$» is never fulfilled, even though x should theoretically converge to zero.

17.3.3. It has occasionally been disputed what the meaning of a relation like «$a = b$» should be, if a is of type **real**, b of type **integer** (see H. C. THACHER, ALGOL Bulletin Nr. 18, item 18.3.1). There is no problem, however, since evaluation of a relation is essentially a comparison between values, thus the above relation produces the value **true** if and only if the current value of a equals the current value of b.

§ 18. Simple Boolean Expressions

Boolean expressions serve to describe calculations with logical values. They may be *simple* or *conditional*. Conditional Boolean expressions are treated later in § 19, while here we deal only with simple Boolean expressions.

18.1. Examples

«$c \wedge d$»	(c and d)		
«$a = 1 \vee b > -1$»	($a = 1$ or $b > -1$)		
«$\neg p$»	(not p)		These five are
«**true**»		These four	at the same
«$x = 2$»		are also	time also
«$(a \equiv \neg p)$»		Boolean	Boolean
«$ff[n\uparrow2]$»		primaries	secondaries

«$g \equiv \neg a \wedge b \wedge (c \supset p = q\uparrow2 \wedge q > p + k) \vee \neg ddcx(u, a \vee b)$»

«$\neg a + b = c \vee d[k+7] \wedge (p \vee q \supset decide(n+1, x=0, roda))$»

Note. According to the syntax, logical constants, simple or subscripted variables of type **Boolean**, relations and Boolean type function designators are already complete Boolean expressions.

18.2. Syntax

A *Boolean primary* is either a logical constant, a (Boolean type) simple or subscripted variable, a relation, a (Boolean type) function designator, or a Boolean expression in parentheses. A *Boolean secondary* is either

just a Boolean primary or a Boolean primary preceded by the operator
« ¬ » (negation). Finally, a *simple Boolean expression* is a sequence of
Boolean secondaries separated from each other by binary Boolean operators (∨, ∧, ⊃, ≡):

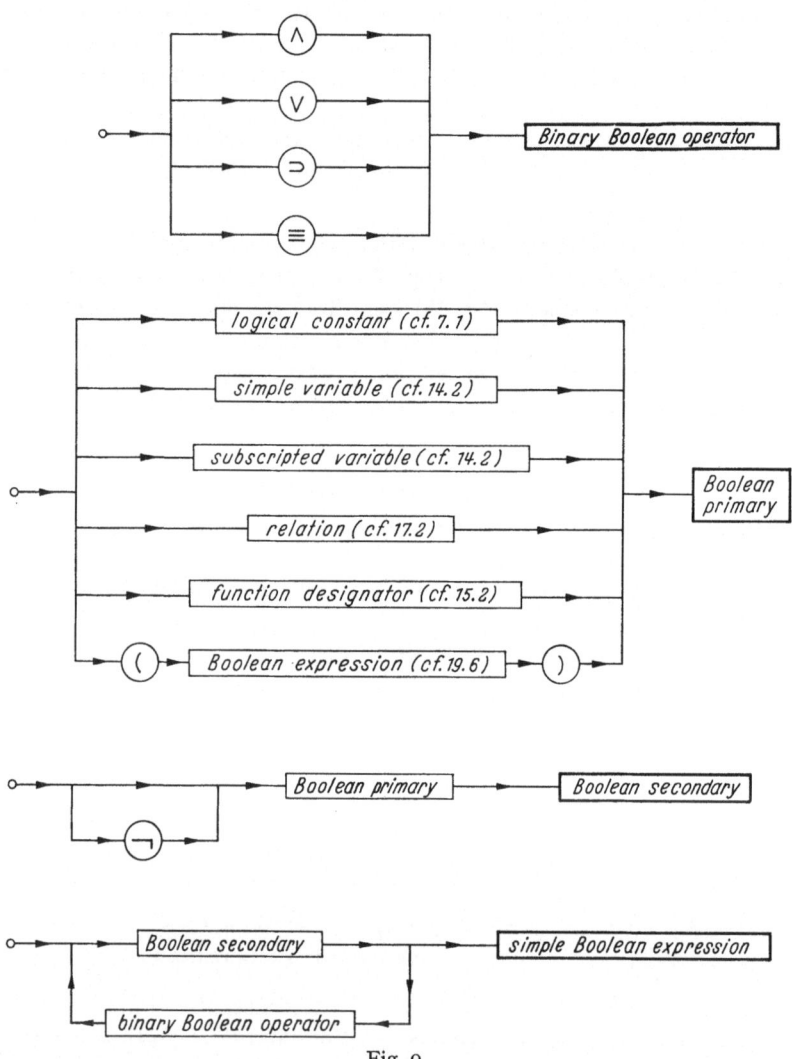

Fig. 9

18.3. Semantics

18.3.1. A simple Boolean expression, when encountered during the execution of an ALGOL program, computes a logical value according to the
following rules:

a) The Boolean primaries of the expression are evaluated independently and their values are used for further evaluation. Arithmetic and Boolean expressions occurring as constituents of the Boolean primaries again are evaluated in accordance with the corresponding rules.

b) Outside the Boolean primaries the following *precedence rules* apply:

First: $\neg$ (negation)
Second: $\wedge$ (conjunction)
Third: $\vee$ (disjunction)
Fourth: $\supset$ (implication)
Fifth: $\equiv$ (equivalence).

c) Except for b), precedence of Boolean operators goes from *left to right*.

The precise content of these rules may be expressed by the following

18.3.2. *Equivalence rule for Boolean expressions:* Let

$$\text{«}S_0\ O_1\ S_1\ O_2\ S_2\ \ldots\ O_n\ S_n\text{»}$$

be a simple Boolean expression, where the O's denote binary Boolean operators and the S's are Boolean secondaries. In this expression additional parentheses can be placed as follows without changing the meaning:

First: All Boolean primaries which are relations are enclosed by parentheses, and all Boolean secondaries which have the form $\text{«}\neg P\text{»}$, are replaced by $\text{«}(\neg P)\text{»}$.

Second: If O_k is an operator which in the sense of 18.3.1,b has precedence over O_{k-1} (or if $k=1$), and O_{k+1} does not have precedence over O_k (or if $k=n$), then the above expression is defined to be equivalent to

$$\text{«}S_0\ O_1\ \ldots\ O_{k-1}\ (S_{k-1}\ O_k\ S_k)\ O_{k+1}\ \ldots\ O_n\ S_n\text{»}.$$

Iterated application of these rules permits parenthesizing a Boolean expression to such an extent that the order in which the operation should be carried out becomes obvious. Treating the last of the examples 18.1 in this way yields

$$\text{«}((\neg(a+b=c))\vee(d[k+7]\wedge((p\vee q)\supset \textit{decide}(n+1,(x=0),\textit{roda}))))\text{»},$$

$$\begin{array}{ccccccc} | & & | & & | & | & | & | \\ 5 & & 6 & & 4 & 1 & 3 & 2 \end{array}$$

which gives a possible order of evaluation by the enumeration below the expression. Likewise

$$\text{«}a\vee b\wedge\neg c\vee d\wedge e\vee\neg f\wedge g\vee h\wedge\neg i\text{»}$$

is transformed into

$$« (((((a \lor (b \land (\neg c))) \lor (d \land e)) \lor ((\neg f) \land g)) \lor (h \land (\neg i))) ».$$

| | | | | | | | | | | |
3 2 1 5 4 8 6 7 11 10 9

18.3.3. *Execution of single logical operations.* The outcome of the operations $\neg a$, $a \lor b$, $a \land b$, $a \supset b$, $a \equiv b$, where a and b represent values of type **Boolean**, is defined by the following table:

Value of a	true	false	true	false
Value of b	true	true	false	false
$\neg a$	false	true		
$a \land b$	true	false	false	false
$a \lor b$	true	true	true	false
$a \supset b$	true	true	false	true
$a \equiv b$	true	false	false	true

18.3.4. *Types.* Of course Boolean operations are defined only if the operands are of type **Boolean**. However, variables and function designators of type **real** or **integer**, as well as numerical constants may still occur in Boolean expressions, namely as subscripts, as actual parameters of function designators and in relations.

§ 19. Conditional Expressions

A conditional (arithmetic, Boolean) expression may be thought of as a device for choosing one of a given set of simple (arithmetic, Boolean) expressions. This choice is made at evaluation time depending on certain logical values.

19.1. Examples

19.1.1. *Conditional arithmetic expressions.*

«**if** $x > 0$ **then** 1 **else if** $x < 0$ **then** -1 **else** 0».

Here a choice is made between the three expressions «1», «-1», «0», the first being evaluated for positive, the second for negative, the third for vanishing x. The whole expression is therefore equivalent to the standard function designator «$sign(x)$».

«**if** $abs(x) \leqq 0.25$ **then** $4 \times x$ **else if** $x > 0$ **then** $2 - 4 \times x$ **else** $-2 - 4 \times x$».

This example chooses between three linear functions in such a way that on the whole the following function is represented:

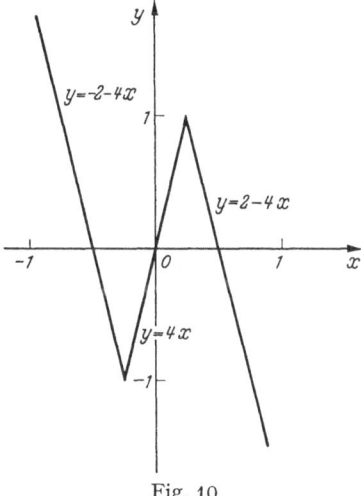

Fig. 10

Thus conditional expressions are, among other things, an excellent instrument for expressing discontinuous functions, but also for interval-wise approximation of analytic functions, e.g.

«**if** $t \geqq 0$ **then** $cos\,(sqrt\,(t))$
  else
    if $t < -0.1$ **then** $(exp\,(sqrt\,(-t)) + exp\,(-\,sqrt\,(-t)))/2$
    else
      $1 - t \times (1 - t \times (1 - t \times (1 - t/56)/30)/12)/2$».

(This computes for any real t the value of the entire function $cos\,(\sqrt{t})$ with an accuracy adequate for a computer with a ten decimal digit mantissa).

19.1.2. *Conditional Boolean expressions.*

«**if** $x = 0$ **then** $z = 0$ **else** $x > 0$».

If $x = 0$, the Boolean expression «$z = 0$» is picked up and evaluated, but if $x \neq 0$, then the logical value of «$x > 0$» is taken as the value of the whole expression. On the whole the meaning of this expression is described by the following truth-table:

	$x = 0$	$x > 0$	$x < 0$
$z = 0$	true	true	false
$z \neq 0$	false	true	false

19.2. Syntax of conditional arithmetic expressions

19.2.1. The conditional arithmetic expression has the syntactic form

$$\text{«if } B_1 \text{ then } E_1 \text{ else if } B_2 \text{ then } E_2 \text{ else} \ldots$$
$$\ldots \text{ else if } B_{n-1} \text{ then } E_{n-1} \text{ else } E_n \text{»,} \tag{1}$$

where the B's represent arbitrary Boolean expressions[1], and the E's represent *simple* arithmetic expressions.

19.2.2. In the RAR, § 3.3.1, the conditional arithmetic expression was defined as

$$\text{«if } B_1 \text{ then } E_1 \text{ else } C_1 \text{»,} \tag{2}$$

with C_1 representing an arbitrary arithmetic expression, while B_1, E_1 have the same meaning as in (1). Since C_1 may again be conditional, (2) is simply a recursive modification of (1), with C_1 standing in place of «if B_2 then E_2 else if ... else E_n».

19.2.3. The syntactic construction «if B then» is called an *if-clause*, while the E's are the *alternatives* of the conditional expression.

19.2.4. *Syntactic diagram*:

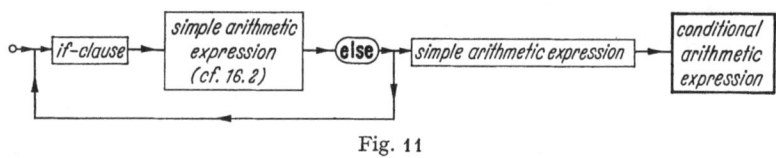

Fig. 11

19.3. Semantics

19.3.1. The value of a conditional arithmetic expression written in the form (2) above is defined as follows:

a) Evaluate the Boolean expression B_1.

b) If B_1 has the value **true**, the desired value of (2) is the result of evaluation of the expression E_1.

c) If B_1 has the value **false**, then the value of the expression (2) is defined to be the value of the expression C_1.

---

[1] In principle, arbitrary Boolean expressions, hence also conditional ones (cf. 19.4), are allowed here, but for reasons of readability it is recommended to use only simple Boolean expressions in if-clauses. This is easily achieved, namely by enclosing conditional Boolean expressions in parentheses.

19.3.2. In view of the fact that C_1 may again be conditional, the evaluation of a conditional expression is described in terms of the more extensive form (1) as follows:

d) Evaluate, in the order from left to right, the Boolean expressions B_1, B_2, ..., etc., until one is found, B_k say, which has the value **true**. Then the value of (1) is defined as the value of E_k. However, if none of the B's has the value **true**, then the value of the last alternative E_n is taken as the value of (1).

Note. After the first Boolean expression B_k having the value **true** has been found, no further B's are evaluated, i.e. the selection of the alternative E_k is independent of the fact that some of the later Boolean expressions B_{k+1}, ..., B_{n-1} might also have the value **true**.

19.4. Conditional Boolean expressions

19.4.1. Syntax and Semantics of these are defined entirely analogously to those of the conditional arithmetic expressions, only that now the alternatives are simple *Boolean* expressions:

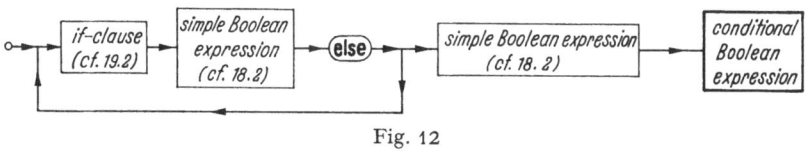

Fig. 12

19.4.2. Conditional Boolean expressions can often be transformed into equivalent simple Boolean expressions. Indeed, the construction

«**if** B_1 **then** B_{11} **else if** B_2 **then** B_{22} **else if** B_3 **then** B_{33} **else** B_{44}»

(where all B's represent simple Boolean expressions) is equivalent to

$$«(B_1) \wedge (B_{11}) \vee (\neg B_1) \wedge (B_2) \wedge (B_{22}) \vee$$
$$(\neg B_1) \wedge (\neg B_2) \wedge (B_3) \wedge (B_{33}) \vee (\neg B_1) \wedge (\neg B_2) \wedge (\neg B_3) \wedge (B_{44})»,$$

provided all B's have well-defined values.

However, if some of the B's are undefined, this equivalence does not hold. As an example, if v is an array with defined components $v[1]$, $v[2]$, ..., $v[n]$, and *kappa* is a Boolean variable, then

«**if** $k = 0$ **then** **true** **else** $kappa \wedge v[k] \neq 0$»

is well-defined for all $k = 0, 1, 2, ..., n$, whereas this is not so for

$$«(k=0) \wedge (\textbf{true}) \vee (\neg k=0) \wedge (kappa \wedge v[k] \neq 0)», \quad \text{(or simplified :)}$$
$$«k = 0 \vee k \neq 0 \wedge kappa \wedge v[k] \neq 0».$$

19.5. Influence of types

The alternatives of a conditional expression must be either all arithmetic or all Boolean. On the other hand, some of the alternatives of a conditional arithmetic expression may be of type **real**, others may be of type **integer**, i.e. these two types may be mixed. However, if this is done, the following rule — a consequence of SR, item 3.3.4 — must be observed:

The value of a conditional arithmetic expression is of type **integer** if and only if all its alternatives are expressions which always produce values of type **integer**[1].

As an example, the expression (n declared **integer**, x declared **real**)

$$\text{«if } n = 1 \wedge c \geqq 0 \text{ then } 13 \text{ else } n{\uparrow}x \text{»}$$

is a real type expression since the expression «$n{\uparrow}x$» is of type **real** (cf. 16.3.3). Therefore, if the first alternative is chosen, the expression has the value 13.0 of type **real**.

19.6. Syntax of general expressions

With the previous definitions we can now define the classes *arithmetic expression, Boolean expression, expression:*

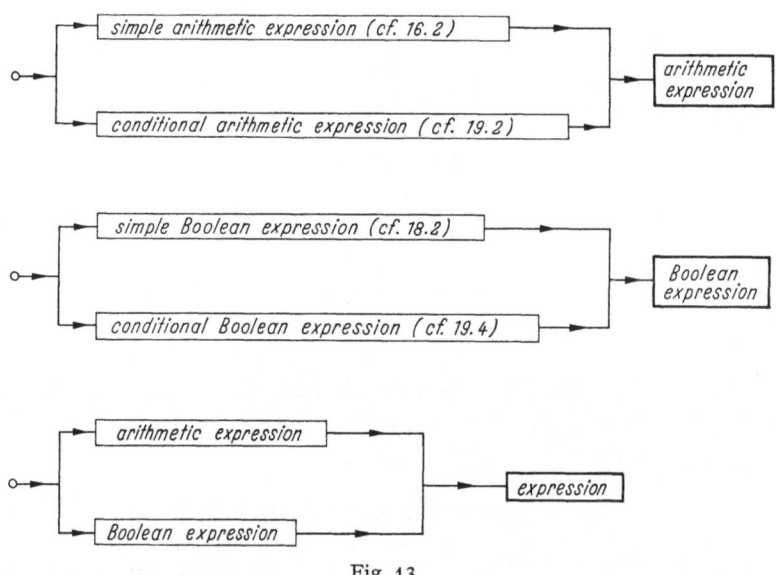

Fig. 13

---

[1] The purpose of this rule is that the type of value of an arithmetic expression can be determined at compilation time, and thus dynamic handling of types becomes unnecessary.

19.7. Further examples involving conditional expressions

19.7.1. $\qquad$ «$x + ($**if** $t > t1$ **then** 1 **else** $-1)/x$».

We recall that a conditional arithmetic expression cannot be used directly as a primary in a larger expression, but must for this purpose be enclosed in parentheses (the sequence «$+$ **if**» is always illegal).

19.7.2. Selection of a component of an array with safeguards against exceeding the array bounds:

«$a[$**if** $k > n$ **then** n **else if** $k < 1$ **then** 1 **else** $k]$».

19.7.3. Where conditional expressions are intended as comparands of a relation or as alternatives of a conditional expression, they must again be enclosed in parentheses:

«**if** (**if** u **then** x **else** y) > 0 **then** (**if** $z = 0$ **then** $x + y$ **else** $x - y$) **else** $x \times y$».

$\underbrace{\qquad\qquad}_{B_1} \qquad \underbrace{\qquad\qquad}_{E_1} \qquad \underbrace{\quad}_{E_2}$

Incidentally, the following is an equivalent form of this expression:

«**if** $(u \wedge x > 0 \vee \neg u \wedge y > 0) \wedge z = 0$ **then** $x + y$
else if $u \wedge x > 0 \vee \neg u \wedge y > 0$ **then** $x - y$
else $x \times y$».

19.7.4. $\quad$ «**if** $a > 0$ **then** $x \neq y$ **else if** $a = 0$ **then** $x < y + 1$ **else** $x > y - 1$».

This is a conditional Boolean expression in which all occurring Boolean elements are relations. It is equivalent to the following simple Boolean expression (the parentheses are not actually needed but are placed for the sake of readability):

«$(a > 0) \wedge (x \neq y) \vee (a = 0) \wedge (x < y + 1) \vee (a < 0) \wedge (x > y - 1)$».

19.7.5. $\qquad$ «**if if** a **then** b **else** c **then** d **else** e»,

(where a, b, c, d, e are Boolean variables). Here the Boolean expression in the if-clause is itself conditional, which is allowed but not recommended. A more readable form is

«**if** (**if** a **then** b **else** c) **then** d **else** e»,

but with the rules of 19.4.2 it could also be transformed into

«$(a \wedge b \vee \neg a \wedge c) \wedge d \vee \neg (a \wedge b \vee \neg a \wedge c) \wedge e$».

19.7.6. The conditional (integer type) arithmetic expression

«**if** $a = 0$ **then** (**if** $b > 2$ **then** (**if** $c < 1$ **then** 1 **else** 2) **else if** $c = d$ **then** 3 **else** 4) **else if** $d > 0$ **then** 5 **else if** $abs(c) < 1$ **then** 6 **else if** $abs(c) = 1$ **then** 7 **else** 8» gives an exact picture of the following tree, insofar as it computes that exit which is used for given values of a, b, c, d:

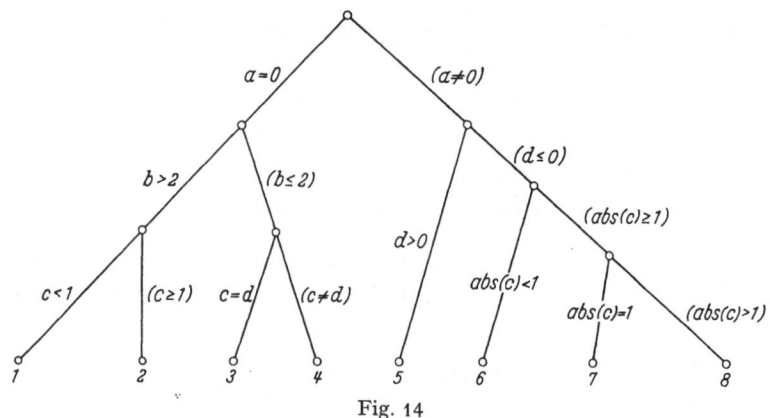

Fig. 14

§ 20. Subscript Expressions

Whenever the syntax requires a subscript expression, this simply means that in principle only an integer-valued arithmetic expression would be meaningful at that position. However, since this would be a great disadvantage in computing practice and partly because there is no syntactic criterion for integer-valuedness of an expression, arbitrary arithmetic expressions are allowed as subscript expressions, but their values are automatically rounded to the nearest integer.

20.1. Syntax

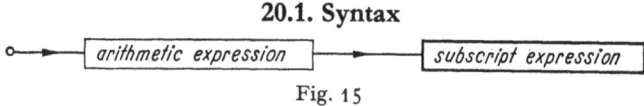

Fig. 15

i.e. a subscript expression is just an arithmetic expression.

20.2. Semantics (Rounding rule for subscript expressions)

Whenever a subscript expression (i.e. an arithmetic expression standing in a position where the syntax requires a subscript expression) is encountered, this expression is first evaluated in accordance with the rules for arithmetic expressions. The value thus obtained is then rounded to the nearest integer and converted to type **integer**. This rounded value is taken as the value of the subscript expression.

20.3. On the use of subscript expressions

20.3.1. Subscript expressions are used in positions where strictly integer values are required, i.e. as subscripts, as actual parameters corresponding to integer type formal parameters (as far as they are called by *value*, cf. 44.6), and on the right side of assignment statements where the left side variables are of type **integer**. It was originally intended that only integer type expressions should be allowed in such places, but it was felt that this would be an impractical restriction. Therefore the rounding rule has been adopted in order to allow in such positions also **real** type arithmetic expressions which theoretically should produce integer values but which in reality contain small deviations caused by roundoff errors (e.g. 178.99999872, -3.000001123, $_{10}-9$).

20.3.2. An example of such an expression is «$n \times (n-1) \times (n-2)/6$», where n is declared **integer**. We know that the resulting value should always be an integer, but since the result of a division is of type **real**, it may deviate by a small amount from an exact integer. A slightly different situation arises with the expression

«$1.1547005 \times sin\,(1.047197551 \times k)$»   ($k$ being declared **integer**),

whose values are (for $k=0, 1, 2, \ldots$): $0, 1, 1, 0, -1, -1, 0, \ldots$.

20.3.3. On the other hand it should be kept in mind that the rounding rule is only intended as a countermeasure in cases where roundoff errors cause deviations from results which theoretically should be integers. It should not be used otherwise. Therefore an expression like

«$3.141592653589 \times k$»   (k declared **integer**),

the value of which is not generally close to an integer, should not be used as subscript expression, though of course the rules of ALGOL would allow it. In such cases it is recommended to achieve the rounding by means of the standard function *entier*.

Chapter IV

Statements

In ALGOL the statements are the units of operation, i.e. the smallest syntactic objects which define closed subcomputations. Various kinds of statements are in use, namely:

The dummy statement: « », i.e. empty space.
Assignment statements, e.g.: «$y := a+b-c[k]/phi$».
Goto-statements, e.g.: «**goto** *arica*».
Procedure statements, e.g.: «*gauss* (a, b, n) *res* : (x)».
For-statements, e.g.: «**for** $k := 1$ **step** 1 **until** n **do** $v[k] := 0$».
Conditional statements, e.g.: «**if** $x > y$ **then** $z := sqrt(x-y)$ **else**
$$y := 1 ».$$

In addition we have the possibility of grouping statements together to compound statements and blocks; these are again considered as statements in all respects.

§ 21. Assignment Statements

An assignment statement serves to assign a computed value (value of an expression) to a simple or subscripted variable in order to preserve that value for later use. This value is then associated with the variable until a further assignment to the same variable overwrites it with another value. More generally, a single computed value may be assigned to several variables simultaneously.

21.1. Examples
«$a := 1$»
«$rec := a \wedge b \vee c \wedge d$»
«$v[k] := arctan((a-b)/(2 \times c))$»
«$dq := rx :=$ **false**».

In the last example, the logical value **false** is assigned to the Boolean variables dq and rx, while in the third example an arithmetic expression is evaluated and its value is assigned to the k-th component of an array v.

21.2. Syntax

An assignment statement consists of an arbitrary sequence of (simple or subscripted) variables, each one followed by an assignment symbol

«:= », the whole being followed by an expression. It therefore has the
syntactic form

$$« V := V := \dots V := E »,$$

where the V's denote variables which are called the *assignment variables*,
and E stands for an arbitrary expression. The most frequently used
special case with only one assignment variable, i.e.

$$« V := E »,$$

is sometimes referred to as a *simple assignment statement*, in contrast to
the more general *multiple assignment statement*.

Syntactic diagram:

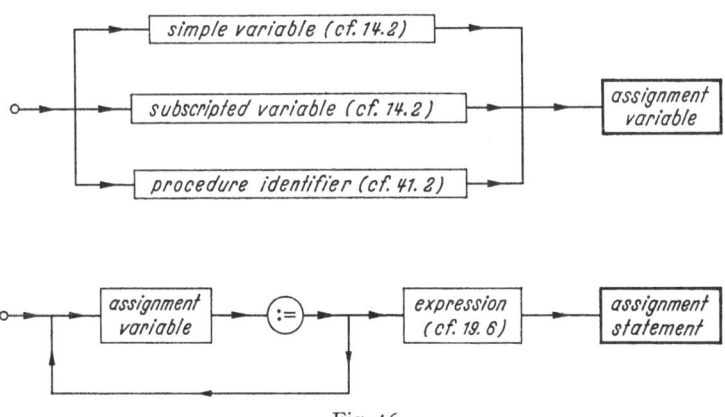

Fig. 16

Note. Only within the body of function procedure I may the procedure
identifier I occur as assignment variable.

21.3. Semantics

21.3.1. If an assignment statement

$$« V_1 := V_2 := \dots V_n := E »$$

is encountered during the execution of a program, the following actions
take place:

a) If subscripted variables occur among the assignment variables,
their subscripts are evaluated first.

b) The expression E is evaluated.

c) The value of E is assigned to all variables V_i.

Though b) and c) will usually suffice to define the effect of an assignment
statement, the full rule is necessary in order to guarantee an unambiguous

result in such cases as

$$«k := a[k] := k+1».$$

Indeed, if this statement is entered with $k=2$, then the value 3 is assigned to $a[2]$, since the subscript 2 is determined *before* k is changed to 3.

21.3.2. *Restrictions.* The fact that simple variables and arrays have scopes and that for subscripted variables the values of the subscripts must lie within the respective subscript bounds as prescribed by the corresponding array declarations, this fact also has certain consequences for assignment statements:

> The effect of an assignment statement is undefined if any one of the assignment variables is nonexistent (in the sense of 42.2) at the location of the assignment statement.

21.4. Influence of types

21.4.1. All assignment variables of an assignment statement must be of the same declared type, i.e. either all **real**, all **integer**, or all **Boolean**. Furthermore, this type must be *compatible* with the type of the expression E on the right side:

a) If the assignment variables are all **Boolean**, then E must be a Boolean expression.

b) If the assignment variables are all **real** or all **integer**, then E must be an arithmetic expression.

21.4.2. In case b) above, the type of the value of E may differ from the type of the assignment variables. However, since only values of the type of the V_i can be assigned to the V_i, the following actions take place (if needed):

a) If the assignment variables are of type **integer**, then E is evaluated as a *subscript expression*, i.e. its value is rounded to the nearest integer and converted to type **integer** before the assignment takes place.

b) If the assignment variables are of type **real**, but the value of E is of type **integer**, then the value of E is converted to real type without changing its value.

21.4.3. Thus examples such as (n, k being declared **integer**, x, y, z declared **real**):

$$«n := (1.618033988{\uparrow}k - (-0.618033988){\uparrow}k)/2.236067977»,\quad \text{and}$$
$$«x := y := z := 3»$$

are meaningful; the first example assigns (for not too large k) the k-th Fibonacci number to n. The second gives the variables x, y, z the value

3.0 of type **real**. On the other hand

   « **real** t ; **integer** k ; $t := k := 0$»

is illegal according to 21.4.1.

§ 22. Sequences of Statements

In an ALGOL program the statements are written one after the other, usually in the order in which they should be executed, and separated from each other by semicolons.

22.1. Examples

22.1.1. The following sequence of statements describes the computation of the rotation angle of a Jacobi rotation in the p, q-plane, $a[i, j]$ being the elements of the matrix to be rotated:

«*theta* := $(a[q, q] - a[p, p])/(2 \times a[p, q])$;
 t := (**if** *theta* > 0 **then** 1 **else** $- 1)/(abs\,(theta) + sqrt\,(1 + theta{\uparrow}2))$;
 c := $1/sqrt\,(1 + t{\uparrow}2)$;
 s := $t \times c$».

The resulting values c, s are the nontrivial elements of the orthogonal rotation matrix U which annihilates the p, q-element of $U^T A U$.

22.1.2.   «*denom* := $a \times e - b \times d$;
       x := $(c \times e - b \times f)/denom$;
       y := $(a \times f - d \times c)/denom$»

describes the solution of the linear equations

$$ax + by = c, \qquad dx + ey = f.$$

The reader should be aware, however, that these examples are far from being complete programs; indeed, the latter must fulfill a number of additional requirements, for which see § 43.

22.2. Syntax

Statements are written in juxtaposition and separated from each other by semicolons (see Fig. 17).

22.3. Semantics

Except for interruptions, omissions and repetitions which are caused by goto-, conditional and for-statements respectively, the statements of an ALGOL program are executed in the order in which they are written down. That is, after the execution of one statement has been completed,

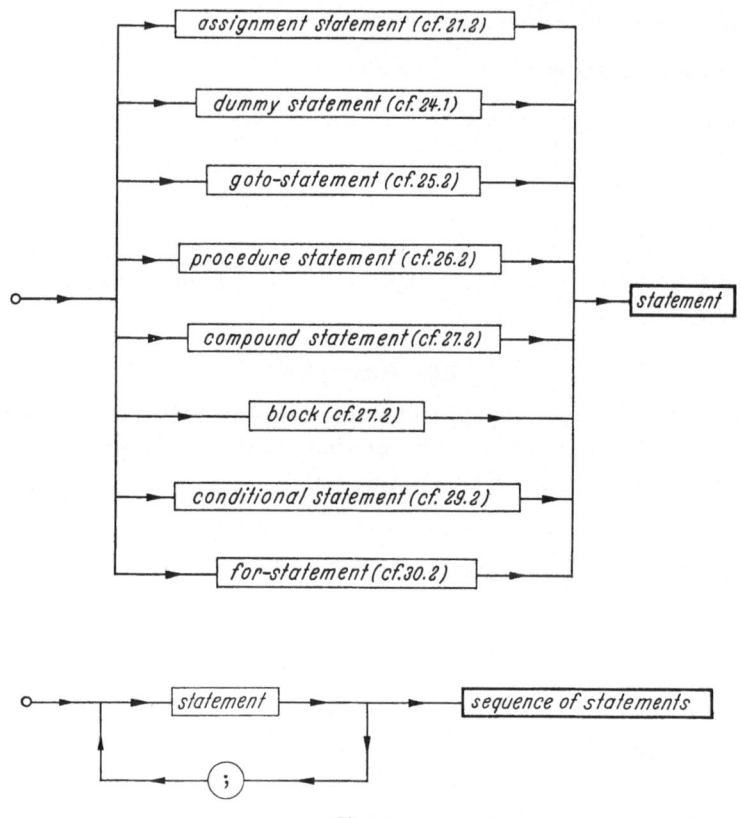

Fig. 17

the statement after the following semicolon comes into action. It is explicitly understood that the execution of one statement does not begin before the execution of the preceding statement has been completed.

§ 23. Labelled Statements

Any statement of an ALGOL program may be furnished with a label (cf. § 11); the two together form a *labelled statement*. Such a label may be placed for explanatory purposes or for marking the destination of a jump.

23.1. Examples

«*arica*: $v[k] := arctan\,((a - b)/(2 \times c))$»
«*jump*: **goto** *arica*»
«*k137*: **for** $k := 1$ **step** 1 **until** n **do** $s := s + v[k]\!\uparrow\!2$»

are labelled assignment, goto-, and for-statements respectively.

«*may* : *label7* : *elim* : *gauss* (*a*, *b*, *n*) *res*: (*x*) »

is a procedure statement with three labels in front of it. According to the rules this is allowed and the whole is again a procedure statement.

23.2. Syntax

23.2.1. If S denotes a statement, and L stands for a label, then

«$L : S$»

is the syntactic form of the labelled statement.

23.2.2. A labelled statement is again considered as a statement of the same kind and therefore has the same properties and is subject to the same restrictions as the corresponding unlabelled statement. Especially, a labelled statement may again be labelled, as shown by the last of the examples 23.1.

23.2.3. *Syntactic diagram*:

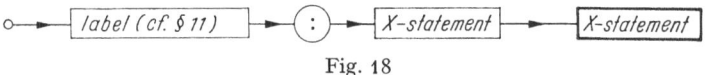

Fig. 18

(valid for $X =$ "*assignment*", "*dummy*", "*goto*", "*procedure*", "*compound*", "*conditional*", "*for*", and mutatis mutandis also for "*block*").

23.2.4. In the following text we shall give all syntactic and semantic definitions for the respective unlabelled statements; the possibility of labelling statements is considered as self-evident and therefore not further mentioned.

23.2.5. *Notations*. The label in front of a statement is called *the label of the statement*, and the statement obtained by depriving the labelled statement of all its labels is called the *unlabelled tail*.

Furthermore, in our text the label in front of a statement will often be used as the name of the statement such that e.g. the first of the statements in 23.1 above would be referred to as "the statement *arica*". It must be pointed out that this is used only for explanatory purposes and has no bearing on the execution of a program.

23.3. Semantics

A labelled statement, whenever encountered during execution of an ALGOL program, has exactly the same effect as the unlabelled tail would have. As a consequence all semantic rules for statements also apply automatically to the respective labelled statements. Indeed, a label has a computational purpose only insofar as a goto-statement may cause a jump from another place in the program to that label.

§ 24. The Dummy Statement

The dummy statement is a statement with no effect whatsoever. It may be used at places where syntactically a statement is required but no effect is wanted.

24.1. Syntax

The dummy statement consists of blank space.

24.2. Semantics

The dummy statement has no effect.

24.3. Examples

A dummy statement may occur through one of the following syntactic combinations:

«; ;», «**begin** ;», «**else** ;», «; **end**», «**begin end**», «**then** ;», «**then else**», «**do** ;», «**then end**», «**do end**».

24.4. Applications

Like all other statements the dummy statement may also be labelled. This permits placing a label where it otherwise would not be allowed, e.g. in front of the symbol «**end**»:

«; *may* : **end**».

Indeed, this combination is used frequently for jumping to the very end of a compound statement, as will be demonstrated by some of the examples in the following chapters.

It should be recognized, however, that the label *may* is not attached to the symbol «**end**», which would be syntactically impossible, but to the dummy statement (blank space) between the semicolon and «**end**». As a consequence, the semicolon in front of the label is indispensable.

§ 25. Goto-Statements

Goto-statements serve to interrupt the normal order of execution of a program by a jump to a specified place.

25.1. Examples

«**goto** *m17*»,
«*jump* : **goto** *arica*»,
«**goto** *ammon* [$k + 17$]».

25.2. Syntax

A goto-statement has one of the syntactic forms[1]

«**goto** L»,   where L denotes an arbitrary label, and

«**goto** $W[E]$», where W is a *switch identifier* (name of a switch), and E is a subscript expression. The construction «$W[E]$» is called a *switch designator* (cf. § 40).

Syntactic diagram:

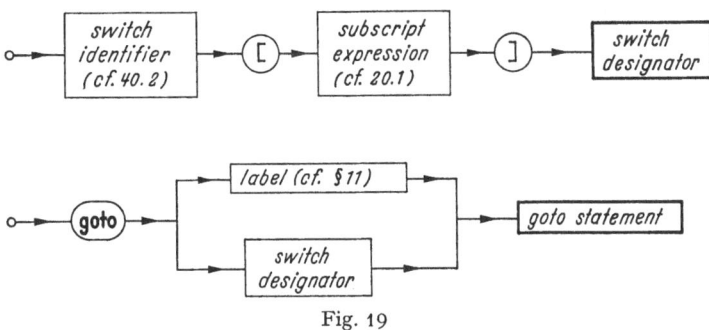

Fig. 19

25.3. Semantics

25.3.1. A goto-statement «**goto** L» causes a jump to that statement S_L in front of which the label L is found as destination label. This has the effect that from now on the statement S_L and all following are executed (again in their natural order) until a further goto-statement is encountered.

25.3.2. A goto-statement «**goto** $W[E]$» is essentially equivalent to «**goto** L_k», where k is the current value of E and L_k is the k-th entry in the switch list of the declaration for the switch W. For more details see § 40, Switch Declarations.

25.3.3. *Restrictions*. In order that the jump to a destination label L be meaningful, the following requirements — derived as special cases of the rules of scopes given in § 42 — must be fulfilled:

For every goto-statement «**goto** L» there must be a unique destination label L such that the smallest *block* or *controlled statement* (whichever is narrower) containing L also contains the goto-statement.

As a consequence, no jump from outside into a block or for-statement is possible, but of course jumps from inside these objects to destinations outside are allowed.

25.3.4. For the restrictions applying to goto-statements of the form «**goto** $W[E]$», see § 40, Switch Declarations.

---

[1] According to the SR, item 3.5.1, no other forms of the goto-statement are possible in SUBSET ALGOL 60.

25.4. Applications

Neglecting for the moment the question of scopes — this will be dealt with in great detail in § 42 — we proceed to some trivial examples of situations that may arise with goto-statements. More realistic examples can only be given in connection with conditional statements (cf. § 28, § 29).

25.4.1.   «**goto** *arica* ;
  comment jump forward, skipping parts of the program ;
  $\vdots$
  arica: $t := t + 1$».

25.4.2.   «*acryl* : *zeta* := *zeta* + 2 ;
  $\vdots$
  goto *acryl* ;
  comment jump backwards, parts of the program are re-
          peated, we obtain a loop ;
  $\vdots$».

25.4.3.   «**begin**
      switch *wernik* := *arica, acryl, m17, larix* ;
      goto *wernik* [*k*] ;
  arica: ; **comment** this for $k = 1$;[1]
      $\vdots$
      goto *common* ;
  acryl: ; **comment** this for $k = 2$;
      $\vdots$
      goto *common* ;
  m17: ; **comment** this for $k = 3$;
      $\vdots$
      goto *common* ;
  larix: ; **comment** this for $k = 4$;
      $\vdots$
  common :
  end».

Here, by virtue of switch *wernik*, the computation follows one of four possible branches of the program depending on the current value of k. Afterwards the common course of the calculation (i.e. the statements which would follow «**end**») is taken up again.

---

[1] The rule that requires a semicolon or «**begin**» in front of «**comment**» forces us to construct a dummy statement «*arica*: ;» if we want to place a comment after a label.

25.5. Closed loops

$log(x)$ can be computed iteratively by a method based on the half-argument formula of the hyperbolic cosine and described by the following piece of program:

$«p := (x + 1/x)/2$;
$\quad q := (x - 1/x)/2$;
iterat:
$\quad p := sqrt(0.5 + 0.5 \times p)$;
$\quad q := q/p$;
goto *iterat*».

Every time the bottom is reached, a jump back to *iterat* occurs with the effect that the last three statements are executed over and over again and thus produce an infinite sequence of pairs p, q (every new pair overwrites its precedessor) where the p's converge to 1 and the q's to $log(x)$. However, since the computation never comes out of this cycle and therefore cannot produce any results, this program is not meaningful. Indeed, *closed loops*, as such never-ending repetitions of the same piece of program are called, must be carefully avoided. As will be shown later, this can be done by making goto-statements conditional; however, as trivial as this may seem, often only a careful analysis, taking the influence of roundoff errors into account, can ensure that a loop is really not closed.

§ 26. Procedure Statements I[1]

A procedure statement serves to initiate execution of an ordinary procedure which either has been declared somewhere in the program (cf. § 41 and Chapter VII) or is one of the standard I/O-procedures described in Chapter VIII.

26.1. Examples

26.1.1. *Unstructurized procedure statements:*

$«gauss : jordan : matinv (75, aa, nores) »$
$«polar»$
$«euler (0, m[k], arctan (0.01), g + 1, equ, f, u) »$
$«remark (x1, 27, 'divergence \sqcup at \sqcup x1=') »$
$«outreal (15, x\uparrow 2) »$.

26.1.2. *Structurized procedure statements:*

$«gauss : jordan : matinv (75) trans:(aa) exit:(nores) »$
$«euler (0, m[k], arctan (0.01), g + 1, equ) trans: (f) res: (u) »$
$«remark (x1) lines: (27) text: ('divergence \sqcup at \sqcup x1=') »$.

---

[1] In this section we describe mainly the syntactic rules for procedure statements; the semantics will be described later in § 45.

26.1.3. By virtue of the corresponding procedure declarations as given in Chapter VII, these examples have the following meaning:

The call of *matinv* inverts a 75×75-matrix, given as an array *aa*, essentially by the Gauss-Jordan method [29] such that after termination of the procedure call the array *aa* contains the computed inverse (the matrix is inverted on the spot!). However, since the diagonal elements are chosen as the pivot elements irrespective of their size, the procedure call may fail because one of these pivots vanishes, in which case a jump to the label *nores* occurs.

The call of *polar* computes, for a point with given cartesian coordinates x, y, its polar coordinates r, *phi*.

The call of *euler* integrates a system of differential equations of order $m[k]$ from $x = 0$ with given initial values $f[1], f[2], \ldots, f[m[k]]$ over $g+1$ steps of length *arctan* (0.01). The system is defined by the declaration for procedure *equ*, which is also given as an example in § 44. The solution is obtained as a two-dimensional array u, where $u[i, j]$ is the j-th component of the solution at the i-th meshpoint.

The call of *remark* prints the text «*divergence at x1=*», followed by the current value of the variable *x1* and 27 blank lines.

According to the definition of the standard I/O procedures, the call of *outreal* outputs the value of $x \uparrow 2$, whereby the output medium and the format are defined by the channel number 15.

26.2. Syntax

26.2.1. The *unstructurized* procedure statement has the syntactic form

$$«I» \text{ or } «I(A, A, \ldots, A)»,$$

where I denotes an arbitrary identifier (the name of the procedure to be called) and the A's are the *actual parameters*, which define the objects upon which the procedure should operate in the present call. The A's may be expressions or identifiers or strings.

26.2.2. *Structurized* procedure statements are obtained by replacing one or more of the commas separating the A's by syntactic objects

$$«)XX \ldots X:(»,$$

where the X represent arbitrary letters. Such a construction is called a *parameter delimiter*.

Note. For the Handbook a special form of the structurized procedure statement is recommended (cf. 44.4.3).

26.2.3. *Syntactic diagram* (see Fig. 20).

26.2.4. *Restrictions.* The syntactic form of a procedure statement is to some extent bound by the corresponding procedure declaration. Indeed, if I stands for the procedure identifier, then the syntactic form must be

«I»,   if the declaration for I has the syntactic form
                               «**procedure** I; S», and
«$I(A_1, A_2, \ldots, A_p)$», if the declaration has the syntactic form
                               «**procedure** $I(P_1, P_2, \ldots, P_p)$; VCS».

In the second case the number of actual parameters must be equal to
the number of formal parameters in the corresponding procedure declaration. Moreover, the k-th formal parameter P_k and the k-th actual parameter A_k (counted from left to right) are said to be *corresponding*, which
in turn implies a number of relations that must hold between A_k and P_k;
these will be stipulated only in § 45.

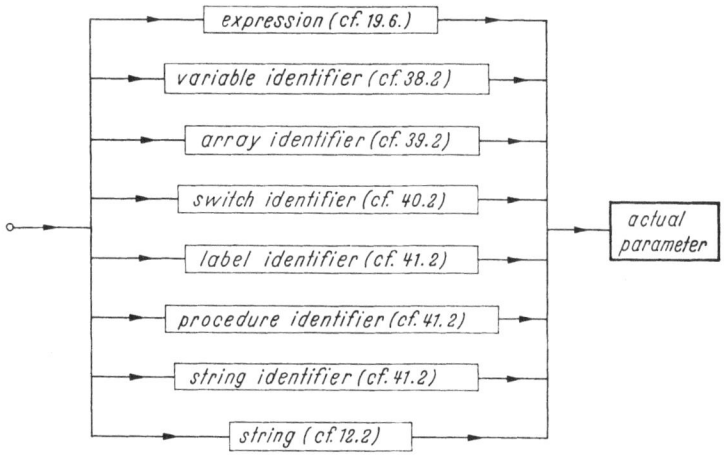

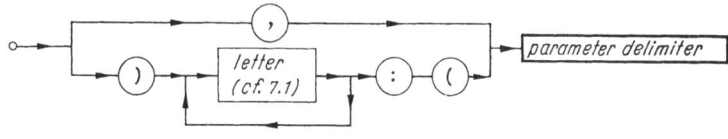

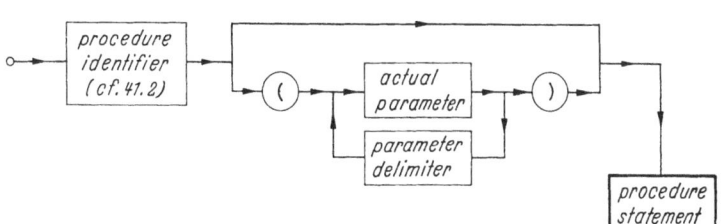

Fig. 20

5*

26.3. Semantics

26.3.1. *Scopes.* All identifiers occurring in a procedure statement must represent (true or formal) quantities which exist at the location of the procedure statement. In other words, a procedure statement must be located within the scope of the procedure which it calls, and also within the scopes of all quantities whose identifiers appear among the actual parameters.

As an example, the procedure statement

$$\text{«} euler\,(0,\; m\,[k],\; arctan\,(0.01),\; g+1,\; equ,\; f,\; u)\,\text{»}$$

must be located within the scope of procedure *euler*, and in the environment of this call, quantities $m, k, arctan$[1], g, equ, f, u must exist.

26.3.2. *Execution of a procedure statement.* A procedure statement causes execution of the corresponding procedure and at the same time defines — through the actual parameters — the quantities and values to be used as operands in that execution. The procedure then performs upon these operands essentially the actions which are prescribed by the procedure declaration for the corresponding formal parameters. For more precise information see Chapter VII.

§ 27. Compound Statements and Blocks

An arbitrary number of consecutive statements may be grouped together into one (compound) statement simply by enclosing them in the word-symbols **«begin»** and **«end»**. This has the effect that certain actions which in principle apply to only one statement (mainly for- and if-clauses) can be extended to operate on several statements simultaneously.

In addition, *declarations* can be inserted after the **«begin»** of a compound statement; in this way we obtain a new element called a *block*.

Compound statements and blocks are themselves statements and therefore may appear in ALGOL programs wherever the rules allow for a statement.

27.1. Examples

27.1.1. The sequence of statements given in 22.1.1 is immediately turned into the following compound statement:

---

[1] *arctan* means here the standard function. This does not mean that the condition that it should exist is superfluous; indeed, the standard functions could also be suppressed by declaring their names for other purposes (cf. 42.2.5).

«begin
    $theta := (a[q,q] - a[p,p])/(2 \times a[p,q])$;
    $t :=$ (**if** $theta > 0$ **then** 1 **else** $-1)/(abs\,(theta) + sqrt\,(1 + theta\uparrow2))$;
    $c := 1/sqrt\,(1 + t\uparrow2)$;
    $s := t \times c$
end».

27.1.2. **«begin**
        $a := c + 1/c - 2$;
        $b := c \times a\uparrow4$;
    mjc: **begin**
           $t := a \times b/c$;
           $v := a/b \times c$
      end *mjc* ;
        $a := a\uparrow1.2$
    end».

This compound statement contains four statements, one of which is itself a (labelled) compound statement

$$«mjc: \textbf{begin } t := a \times b/c ; \; v := a/b \times c \textbf{ end»}.$$

27.1.3. «*block*: **begin**
           real t ;
           $t := (1 + a\uparrow3 + a\uparrow6)/(1 + a\uparrow2 + a\uparrow4)$;
           $a := exp\,(t\uparrow2/2) - a \times cos\,(t) + sqrt\,(t + 1/t)$
        end»

is a labelled block. The declaration at its beginning introduces a new variable t which is existent only within this block and serves there as an auxiliary variable for storing the value of the expression

$$«(1 + a\uparrow3 + a\uparrow6)/(1 + a\uparrow2 + a\uparrow4)»$$

temporarily.

27.2. Syntax

27.2.1. Compound statements and blocks have the syntactic forms

    «**begin** S ; S ; ... ; S **end»**   and
    «**begin** D ; D ; ... ; D ; S ; S ; ... ; S **end»**

respectively, where the S's denote *arbitrary statements* and the D's represent arbitrary *declarations* (cf. Chapter VI).

    The construction «**begin** D; D; ...; D ;», which contains all declarations of a block, is called the *block head*.

    Note. Since compound statements and blocks are again statements, the following are allowed constructions:

«**begin** S ; S ; **begin** D ; D ; S ; S **end** ; **begin** S ; S **end** ; S **end**»
«**begin** D ; S ; **begin** D ; S ; **begin** D ; S ; S **end end end**».

However, it is not allowed to precede a declaration with a statement; indeed, a declaration may only be placed after a **begin** or after another declaration.

27.2.2. *Syntactic diagram:*

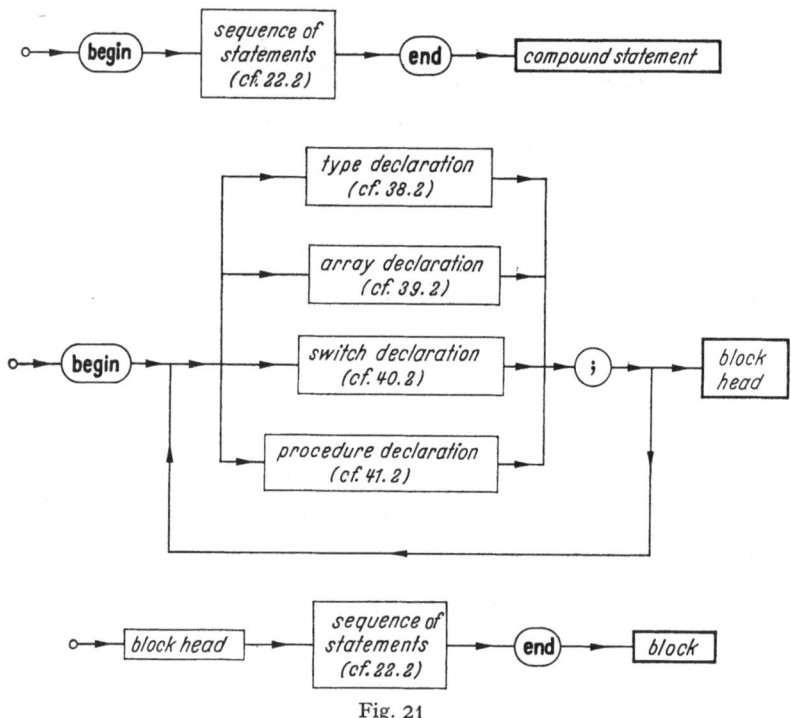

Fig. 21

27.2.3. *Remarks.* Since the semicolon is required only for separating statements, no semicolon is needed in front of the **end** (If we do place one, this introduces a dummy statement but causes no error). However, if a compound statement or block is followed by another statement, a semicolon is needed after the «**end**» (If we forget it, the subsequent statement is considered as a comment, and this is indeed an error).

27.3. Semantics of compound statements[1]

27.3.1. The execution of a compound statement is equivalent to the execution of the sequence of statements contained in it, and this is described in 22.3. However, the following exceptions must be observed:

---

[1] For the semantics of blocks see § 42.

In addition to the *normal* entrance to a compound statement, i.e. the entrance through «**begin**», it may also be entered by a jump from outside to one of the labels contained in it. In that case the compound statement is executed from that label onwards. Moreover, in addition to the *normal* exit from a compound statement, i.e. the exit through «**end**», an exit from a compound statement may also occur by a jump to a label located outside.

27.3.2. It is thus apparent that the «**begin**» and «**end**» of a compound statement have no bearing on the execution of a program, with the exception of those cases where the compound statement is

either a *complete program* (cf. § 43),
or the *controlled statement of a for-statement* (cf. § 30),
or an *alternative of a conditional statement* (cf. § 28 and § 29),
or a *procedure body* (cf. § 44).

For instance, in the compound statement given as example 27.1.2, the «**begin**» and «**end**» of the internal compound statement (labelled *mjc*) could be omitted without the slightest alteration of the operational effect of this piece of program. Whether the outer «**begin**» and «**end**» could also be removed depends upon the environment into which this compound statement will be embedded.

§ 28. The If-Statement

A statement can be made conditional by placing an *if-clause* in front of it. This if-clause states the condition under which the subsequent statement is executed.

If-clause and subsequent statement together are the *if-statement*, which is the simplest kind of conditional statement, while all kinds of statements described in § 21—27 are called *unconditional statements*.

28.1. Examples

28.1.1. «**if** $x = 0$ **then** $x := {}_{10} - 20$».

This obviously means that a vanishing x is replaced by a small non-zero number, e.g. in order to avoid trouble in a later division. If $x \neq 0$, this statement has no effect.

28.1.2. «**if** $k > 0 \wedge k < 5$ **then goto** *wernik* $[k]$».

$\underbrace{ k > 0 \wedge k < 5 }_{\text{if-clause}} \quad \underbrace{ \textbf{goto } wernik[k] }_{\text{goto-statement}}$

It is assumed that *wernik* is the switch declared by the example in 25.4.3. Here, the if-clause serves to prevent the computation from running into the undefined situation mentioned in 40.3, c.

28.1.3. (cf. 22.1.1).

«**if** *rotate* **then**
  begin
    $theta := (a[q,q] - a[p,p])/(2 \times a[p,q])$;
    $t := ($**if** $theta > 0$ **then** 1 **else** $-1)/(abs(theta) + sqrt(1 + theta \uparrow 2))$;
    $c := 1/sqrt(1 + t \uparrow 2)$;
    $s := c \times t$
  end».

Here the whole compound statement is subject to the if-clause, i.e. if the Boolean variable *rotate* has the value **false**, none of the statements is executed. Obviously «**begin**» and «**end**» have an operational meaning in this case, namely they prevent the first of the four statements from being the only one subject to the if-clause.

28.1.4. «**if** $x > 0$ **then** $put : z := $ **if** $y < x$ **then** x **else** y».

           if-clause   labelled assignment statement
                              (with conditional expression on the right side).

28.1.5.

                Conditional Boolean expression

«**if** (**if** $x \geq 0$ **then** $z > cos(sqrt(x))$ **else** $z > coshyp(sqrt(-x))$)) **then** $x := 0$».
                    if-clause   state-
                                                           ment

Assuming that «$coshyp(x)$» is a function designator which computes the hyperbolic cosine of the argument, the condition in the if-clause checks whether the value of z exceeds the value of the entire function $cos \sqrt{x}$.

28.2. Syntax

The if-statement has the syntactic form

«**if** B **then** S_{if}»,

where B represents an arbitrary[1] Boolean expression and S_{if} is any statement whose unlabelled tail does not begin with the symbol «**if**». Where an if-statement is followed by the symbol «**else**», it is always part of an *if-else-statement*, for which see § 29.

---

[1] As in the case of conditional expressions, it is also recommended to use only simple Boolean expressions here.

Syntactic diagram[1]:

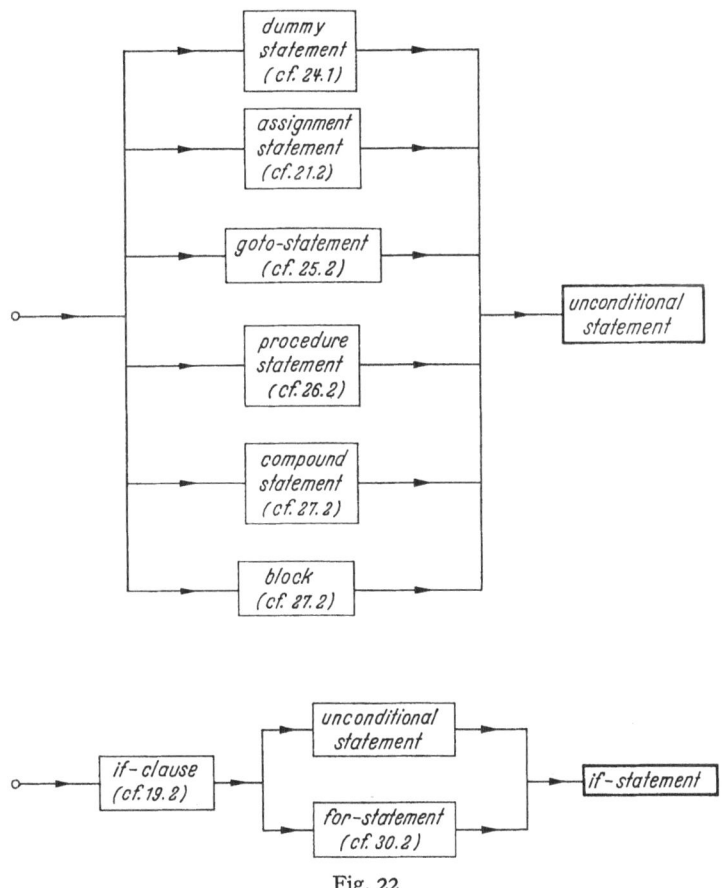

Fig. 22

28.3. Semantics

28.3.1. The execution of an if-statement «**if** B **then** S_{if}» involves the following operations:

a) Evaluation of the Boolean expression B, and
b) Execution of the statement S_{if}, if B has the value **true**, but
c) No further action, if B has the value **false**.

---

[1] The reason why the for-statement is not ranked among the unconditional statements (this is a difference between the original and the revised ALGOL report) is to avoid the ambiguity that could arise in connection with for- and if-else-statements (cf. 30.4).

Thus an if-statement with $B =$ **false** is equivalent to a dummy statement, while for $B =$ **true** it is as if the if-clause were not present[1].

28.3.2. If a jump from outside S_{if} is directed to a destination label which is part of S_{if}, then the if-clause is disregarded. Thus a jump to the label *put* in example 28.1.4 would be allowed and would cause the unconditional execution of the statement «$z := \ldots$», whereupon the subsequent statement is taken up:

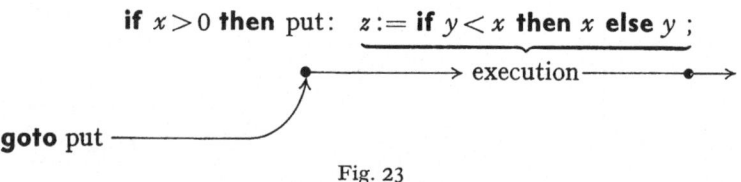

if $x > 0$ **then** put: $z :=$ **if** $y < x$ **then** x **else** y ;

goto put

Fig. 23

28.4. Applications

28.4.1. The historical background of the if-statement is the *conditional jump* as a machine code instruction, and indeed also in ALGOL the corresponding device «**if** ... **then goto** ...» is used frequently, especially for breaking closed loops.

However, the customs in writing programs in machine code need not necessarily be taken over into ALGOL. For instance, in ALGOL the example 28.1.3 is more appropriate than the equivalent form

«**if** $\neg rotate$ **then goto** *over* ;
  $theta := (a[q, q] - a[p, p])/(2 \times a[p, q])$;
  $t := ($**if** $theta > 0$ **then** 1 **else** $- 1)/(abs\,(theta) + sqrt\,(1 + theta \uparrow 2))$;
  $c := 1/sqrt\,(1 + t \uparrow 2)$;
  $s := t \times c$;
over : ... »,

which has a strong scent of machine code.

28.4.2. The *breaking of closed loops* may now be demonstrated: Take for instance the example in 25.5; it can now be modified into

«$p := (x + 1/x)/2$;
  $q := (x - 1/x)/2$;
iterat: $p := sqrt\,(0.5 + 0.5 \times p)$;
  $q := q/p$;
  if $p > 1.000015$ **then goto** *iterat* ;
final: $\log := 3 \times q/(2 + p)$ ».

Obviously the jump back to *iterat* occurs as long as $p > 1.000015$, which, because p converges to 1, will not be true forever. Thus, sooner or later,

---

[1] This is correct only in SUBSET ALGOL 60 because otherwise the evaluation of function designators occurring in the Boolean expression B could produce a side effect (cf. 46.5).

the loop will be discontinued, and the computation will continue with the statement *final*.

An equivalent setup is

$$\ll p := (x + 1/x)/2 \; ;$$
$$q := (x - 1/x)/2 \; ;$$
iterat: **if** $p > 1.000015$ **then**
    begin
        $p := sqrt(0.5 + 0.5 \times p) \; ;$
        $q := q/p \; ;$
        goto *iterat*
    end ;
final: $log := 3 \times q/(2 + p) \gg .$

The only difference between this and the first version is that when (in the second version) for the first time $p \leq 1.000015$, the jump to *iterat* occurs again, but then the if-statement acts as a dummy statement, and therefore its successor — the statement *final* — comes into action.

In devising such a scheme for breaking closed loops, it must be ensured that it works not only with the theoretically expected values, but also with the numbers occurring in the actual computation. In the above examples the roundoff errors might have the effect that the numerically calculated values p forever remain above 1.000015, and then we would again have a closed loop despite our attempts to prevent it. In the above examples the constant 1.000015 (chosen to yield a 10-digit logarithm) is far enough away from 1 to guarantee discontinuation of the loop for computers with at least an eight-digit (decimal) mantissa.

28.4.3. *Warning:* According to the syntax, the sequence

$$\ll \textbf{if } x = 0 \textbf{ then if } a > b \textbf{ then } t := 1 \gg$$

is obviously not allowed. To describe the desired effect in correct ALGOL, we must write instead

$$\ll \textbf{if } x = 0 \textbf{ then begin if } a > b \textbf{ then } t := 1 \textbf{ end} \gg$$

(in this way the statement following the first «**then**» becomes unconditional). We note in passing that as long as the values of a, b cannot be undefined, the same could also be achieved by

$$\ll \textbf{if } x = 0 \land a > b \textbf{ then } t := 1 \gg .$$

However, this latter form may be less efficient in cases where the if-statement is executed frequently and the condition $x = 0$ seldom fulfilled.

28.4.4. It goes without further mention that in the sequence

$$\ll \textbf{if } z \neq 0 \textbf{ then } p := 0 \; ; \textbf{ goto } arica \gg ,$$

only the statement «$p := 0$» is subject to the if-clause. If it is intended that the jump, too, should be conditional, the whole must be rewritten as

«**if** $z \neq 0$ **then begin** $p := 0$; **goto** *arica* **end**».

§ 29. The If-Else-Statement

The if-else-statement is an extension of the if-statement and allows selecting and executing one of several unconditional statements, the latter being called the *alternatives* of the if-else-statement. The selection is made via a number of conditions which, together with the alternatives, constitute the if-else-statement.

The if-else-statements together with the if-statements form the class of *conditional statements*.

29.1. Examples

29.1.1.   «**if** $x > 0$ **then goto** *posi* **else goto** *nega*»

           if-clause   first alter-   second alter-
                     native   native

This statement causes a jump to *posi* if $x > 0$ or a jump to *nega* if $x \leq 0$.

29.1.2. Intervalwise approximation of the Bessel function $J_0(x)$:

```
«if abs (x) < 8 then
  begin
    real t ;
    t := − x↑2/32 ;
    j0 := 0.999 999 99 + t × (7.999 999 99 + t × (15.999 998 78
        + t × (14.222 203 20 + t × (7.111 007 52 + t × (2.275 260 80
        + t × (0.505 177 60 + t × (0.082 021 76 + t × (0.009 950 72
        + t × (0.000 860 16 + t × 0.000 040 96)))))))))
  end
else
    begin
      real t, p0, q0, y ;
      t := − 64/x↑2 ;
      p0 := 0.797 884 56 + t × (0.000 876 54 + t × (0.000 021 57
          + t × 0.000 001 28)) ;
      q0 := 0.012 466 95 + t × (0.000 114 15 + t × (0.000 005 49
          + t × 0.000 000 51)) ;
      y  := x − 0.785 398 163 × sign (x) ;
      j0 := (p0 × cos (y) + q0 × sin (y) × 8/x)/sqrt (abs (x))
    end».
```

Both alternatives of this if-else-statement are blocks; the first block approximates the power series between -8 and $+8$, the second uses asymptotic series for representing the function outside the interval $|x| < 8$. The maximum error is on the order of $_{10}-7$.

29.1.3. Example with more than two alternatives (square root $x+iy$ of a complex number $c = a+ib$):

«$p := sqrt(a{\uparrow}2 + b{\uparrow}2)$;	$\begin{cases}\text{this statement is outside}\\ \text{the conditional statement.}\end{cases}$
if $p=0$ **then** $x := y := 0$	first alternative: $c=0$.
else	
if $a>0$ **then**	
begin	
$x := sqrt((a+p)/2)$;	$\left.\begin{array}{l}\\ \\ \\ \end{array}\right\}$ second alternative: if c is
$y := b/(2 \times x)$	in the right half plane.
end	
else	
begin	
$y := $ **if** $b \geq 0$ **then** $sqrt((p-a)/2)$	third alternative: if c is in
else $-sqrt((p-a)/2)$;	the left half plane or on
$x := b/(2 \times y)$	the imaginary axis, but
end».	not zero.

29.2. Syntax

29.2.1. An if-else-statement has basically the syntactic form

(1)   «**if** B **then** U **else** S»,

where B denotes a Boolean expression, U an *unconditional* statement, i.e. one beginning with neither »**for**« nor »**if**«, and S stands for an *arbitrary statement*.

29.2.2. Since the S in (1) above may itself be an if- or an if-else-statement, we are finally lead to the most general form of the conditional statement, expressed in terms of unconditional and for-statements[1]:

The *open form:*

(2)   «**if** B_1 **then** U_1 **else if** B_2 **then** U_2 **else** ... **else if** B_n **then** S_{if}»

The *closed form:*

(3)   «**if** B_1 **then** U_1 **else if** B_2 **then** ... **else if** B_n **then** U_n **else** S_{if}».

---

[1] As in § 28, S_{if} denotes a statement whose unlabelled tail does not begin with the symbol «**if**», i.e. an unconditional or a for-statement.

29.2.3. *Terminology.* In the syntactic forms (2), (3) above the statements U_j and S_{if} are called the *alternatives* of the if-else-statement. This term is sometimes also used for the simple if-statement, which occurs as a special case of the open form for $n = 1$, although the term may seem somewhat misplaced in this case.

29.2.4. *Syntactic diagram for the if-else-statement:*

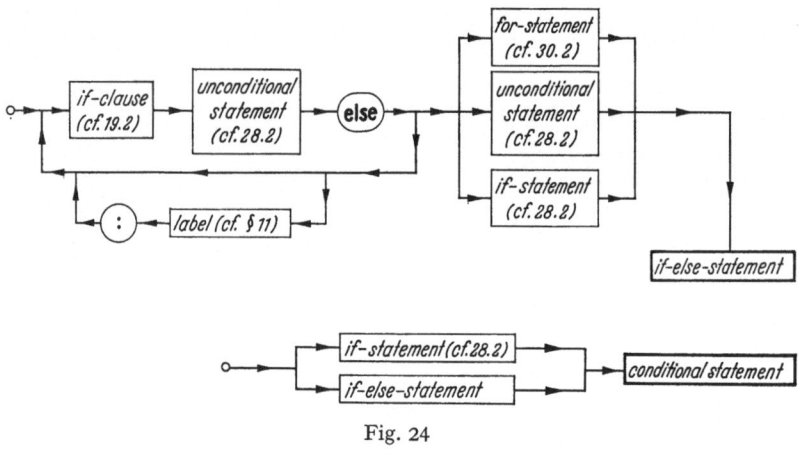

Fig. 24

29.2.5. *Labels in an if-else-statement.* Since in the syntactic representation (1) the statement U as well as S may be labelled, labels may appear in the syntactic representations (2), (3) not only in front of the alternatives U_j, S_{if}, but also in front of any if-clause. Thus

«*p*: **if** $x = 0$ **then** q: $y := 0$ **else** r: **if** $y \neq 0$ **then** s: $x := 2$ **else** t: **goto** f»

$$\underbrace{}_{\text{if-clause}} \quad \underbrace{}_{U} \quad \underbrace{}_{S}$$

$$\underbrace{}$$

would be allowed.

29.2.6. *Warning.* It seems that because of the rule which requires that statements be separated by semicolons, programmers have the habit of automatically placing a semicolon after every statement. Some of these may be superfluous insofar as they just generate dummy statements which do no harm. However, if a semicolon is placed after one of the alternatives (except the last one) of an if-else-statement, e.g.

«**if** $x = 0$ **then goto** *posi* ; **else goto** *nega*»,

this is a syntactical error; in fact, the combination «; **else**» is never correct outside strings.

29.3. Semantics

29.3.1. *The simplified rule.* As long as no jumps into one of the alternatives occur, the effect of an if-else-statement represented in the form (1) can be visualized as follows:

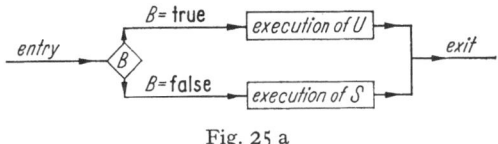

Fig. 25 a

As a consequence of this we immediately obtain an analogous picture defining the effect of an if-else-statement represented by one of the syntactic forms (2) or (3):

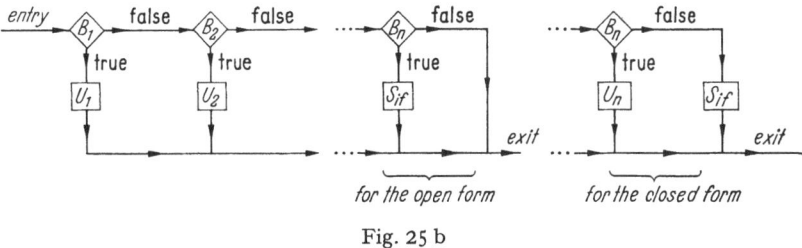

Fig. 25 b

29.3.2. *Equivalence rule for if-else-statements.* Usually the above diagrams will be sufficient, but the most general case requires the following more precise rule:

An if-else-statement «**if** B **then** U **else** S» is defined to be equivalent to the compound statement

  «**begin**
     if B **then begin** U ; **goto** L **end** ;
     S ;
L:
    end»,

in which L denotes a label considered to be different from any other label.

In case S is again an if-else-statement, the same rule is applied again to S, etc. until finally (and after removal of some unnecessary **begin**'s and **end**'s) the following equivalent forms for the syntactic representations (2) and (3) are obtained:

For the open form (2):

>«begin
>>if B_1 then begin U_1 ; goto L end ;
>>if B_2 then begin U_2 ; goto L end ;
>>
>>$\vdots$
>>
>>if B_{n-1} then begin U_{n-1} ; goto L end ;
>>if B_n then S_{if} ;

L:

>>end».

For the closed form (3):

>«begin
>>if B_1 then begin U_1 ; goto L end ;
>>if B_2 then begin U_2 ; goto L end ;
>>
>>$\vdots$
>>
>>if B_n then begin U_n ; goto L end ;
>>S_{if} ;

L:

>>end».

In this way the if-else-statement is expressed entirely in terms of if-statements and therefore the semantics of the former follows now from § 28.

29.3.3. An example for which only the general rule gives the correct answer, is [1]

>«if $x=0$ then adv: $x := y - 1$ first alternative
>else
>r: if $y > 1$ then
>>begin
>>>$y := y + c$;
>>>if $c < 0$ then goto r ;   second alternative
>>>if $c > 1$ then goto adv
>>
>>end
>>else
>>>$polar$».   third alternative

With the equivalence rule above, this transforms into

>«begin
>>if $x=0$ then begin adv: $x := y - 1$; goto $exitt$ end ;
>r: if $y > 1$ then

---

[1] Needless to say we do not recommend such jumping within an if-else-statement, since it tends to disguise the intentions of the programmer and thus may give a skew picture of the computing process.

begin
  begin
    $y := y + c$;
    if $c < 0$ **then goto** r ;
    if $c > 1$ **then goto** adv
  end ;
    goto $exittt$
  end ;
    $polar$;
$exittt$:
  end».

It follows that if the given statement is entered e.g. with $x \neq 0$, $y > 1$, then the second alternative is taken up. If furthermore $c < 0$, then a jump from inside the second alternative back to the label r occurs, which causes a second entry into the second alternative. This repetition continues until finally $y \leq 1$, whereupon the third alternative (procedure statement «$polar$») is executed, after which the execution of the if-else-statement is terminated. If, however, c was > 1, then «**goto** r» is skipped and «**goto** adv» executed instead, which causes a jump to and execution of «$x := y - 1$», after which the execution of the if-else-statement again terminates.

29.3.4. *If-else-versus sequence of if-statements.* Example 25.4.3 could now be rewritten equivalently as

«**if** $k = 1$ **then** $arica$: **begin** ... **end**
  else
    if $k = 2$ **then** $acryl$: **begin** ... **end**
    else
      if $k = 3$ **then** $m17$: **begin** ... **end**
      else
        if $k = 4$ **then** $larix$: **begin** ... **end**».

It would seem that in this example, where the four conditions are disjoint, the **else**'s could just as well be replaced by semicolons, thus splitting the if-else-statement into four if-statements. Often this is true; however, k might undergo changes in one alternative, which could have the effect that one of the later conditions would also be fulfilled, and then the sequence of if-statements would no longer be equivalent to the if-else-statement. In this sense the if-else-statement is safer, insofar as the rules guarantee that always only one of the alternatives is executed.

29.4. Efficiency considerations

Obviously the rules allow arranging if- and if-else-statements in several ways, all of which yield the same effect but may differ widely in efficiency. No rules can be given as to which setup is the most economical, but a few examples may be helpful:

29.4.1. Example 25.4.3 is a more efficient form than 29.3.4 since the latter requires testing of two conditions on the average while the former directly selects the proper label.

29.4.2.   «**if** $x \neq 0 \wedge y > 0$ **then** *alpha*
        else
            if $x \neq 0 \wedge z = 0$ **then** *beta*
            else
                if $x \neq 0 \wedge z > 2$ **then** *gamma*
                else
                      if $x \neq 0$ **then** *delta*»

(*alpha*, *beta*, *gamma*, *delta* being procedures without parameters). If the condition $x \neq 0$ is seldom fulfilled, then it is better to test this condition first, making the other conditions subordinate to it:

«**if** $x \neq 0$ **then**
  begin
    if $y > 0$ **then** *alpha*
    else
      if $z = 0$ **then** *beta*
      else
        if $z > 2$ **then** *gamma*
        else
          delta
  end».

Indeed, in this way $x \neq 0$ is tested only once, and only if the condition holds, are the other conditions tested at all.

An equally economical version is

«**if** $x = 0$ **then**
  else
    if $y > 0$ **then** *alpha*
    else
      if $z = 0$ **then** *beta*
      else
        if $z > 2$ **then** *gamma*
        else
          delta».

We observe that the first alternative is a dummy statement which causes skipping of the whole in case $x \neq 0$ is not true.

§ 30. The For-Statement

A for-statement is a shorthand notation for a loop. It consists of a *controlled statement* and a preceding *for-clause*. The latter defines how often and for what values of the running subscript the controlled statement should be executed.

30.1. Examples

30.1.1. «**for** $i := 1$ **step** 1 **until** n **do** $v[i] := 0$».

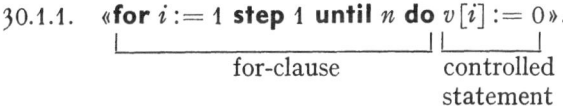

     for-clause   controlled
              statement

This for-statement annihilates all components 1 through n of the vector $\vec{v}$ and thus in a certain sense represents the vector operation $\vec{v} := \vec{0}$.

30.1.2. «$h := 0$;
    for $k := n$ **step** -1 **until** 0 **do** $h := h \times x + a[k]$».

       for-clause   controlled
                 statement

Here the controlled statement «$h := h \times x + a[k]$» is executed once for each of the values $k = n, n-1, n-2, \ldots, 1, 0$ (in that order) and thus by virtue of the initialisation «$h := 0$» computes the value of the polynomial $\sum_{k=0}^{n} a[k] x^k$ by HORNER's rule. Thus, obviously, a running subscript may also run backwards.

30.1.3. Example with a *nested loop* (generation of unit matrix):

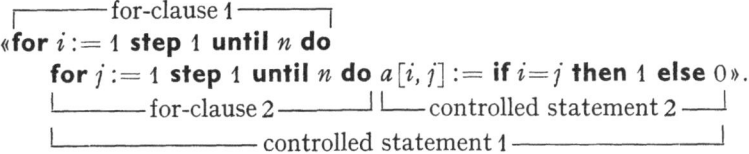

  ┌────── for-clause 1 ──────┐
«**for** $i := 1$ **step** 1 **until** n **do**
   for $j := 1$ **step** 1 **until** n **do** $a[i, j] := $ **if** $i = j$ **then** 1 **else** 0».
   └────── for-clause 2 ──────┘ └── controlled statement 2 ──┘
   └──────────── controlled statement 1 ────────────┘

Controlled statement 1, which is itself a for-statement, is executed for all i, hence controlled statement 2 is excuted for all i *and* j.

30.1.4. Multiplication of matrix a with a vector b:

  «**for** $i := 1$ **step** 1 **until** n **do**
comp: **begin**
    $c[i] := 0$;
    for $j := 1$ **step** 1 **until** n **do** $c[i] := c[i] + a[i, j] \times b[j]$
   end i».

6*

The controlled statement of the for-i-clause is compound statement $comp$, which is therefore executed (as a whole) once for every value $i=1, 2,$ $\ldots, n$. Every execution of $comp$ of course includes complete performance of the for-j-statement.

30.1.5. «**for** $k := 0$ **step** 1 **until** n **do**

$\qquad\qquad kappa :=$ **if** $k=0$ **then true else** $kappa \wedge v[k] \neq 0$»

tests whether all components of an array $v[1:n]$ are nonzero, and only if this is so, does $kappa$ obtain the final value **true**. Note that here 19.4.2 cannot be applied.

30.1.6. Other ways of governing the running subscript (which in this case is not actually a subscript but a real type variable) are demonstrated by the example

«**for** $z := x{\uparrow}2, a[0], -1, 1, z/2$ **while** $z >_{10} -6$ **do begin** ... **end**»

$\qquad$ single expression $\qquad$ while element
$\qquad\qquad$ elements

$\qquad\qquad\qquad$ for-clause $\qquad\qquad\qquad$ controlled
$\qquad\qquad\qquad\qquad\qquad\qquad\qquad\qquad\qquad$ statement

Here z runs through the values $x{\uparrow}2$, $a[0]$, -1, 1, and then 1/2, 1/4, 1/8, etc. until 1/524288 (which is the last one fulfilling the condition $z >_{10} -6$), and for every one of these z the whole compound statement «**begin** ... **end**» is executed.

30.2. Syntax

30.2.1. The for-statement has the syntactic form

«**for** $V := F$ **do** S»,

where V represents a simple variable of real or integer type (called the *controlled variable*), F is the so-called *for-list*, and S is an arbitrary statement (the *controlled statement*). The construction «**for** $V := F$ **do**» is called the *for-clause*.

30.2.2. In most applications, e.g. in the examples 30.1.1-4, the for-list will have the syntactic form «E_1 **step** E_2 **until** E_3», where the E's denote arbitrary arithmetic expressions. It means that the controlled variable runs through a strictly linear sequence

$$E_1, E_1 + E_2, E_1 + 2 \times E_2, \text{ etc. (until } at\ most\ E_3).$$

30.2.3. In the most general case, however, the for-list can be a construction

«$H, H, \ldots, H$»,

whose entries H, called the *for-list elements*, are separated by commas and may have one of the following syntactic forms:

«E»   (single expression element),
«E_1 **step** E_2 **until** E_3»   (step-until element),
«E **while** B»   (while element),

where in all three cases the E's denote arbitrary arithmetic expressions, B represents a Boolean expression, and «**while**», «**until**», and «**step**» are basic symbols of the language.

30.2.4. *Syntactic diagram*:

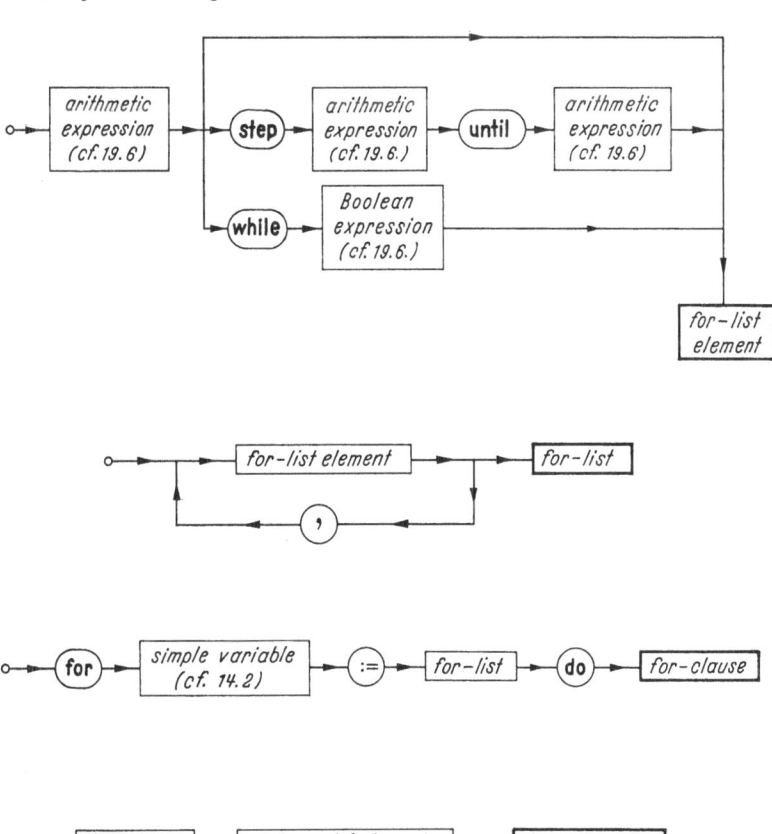

Fig. 26

30.3. Semantics

30.3.1. *Static for-statements*. A for-statement is called *static* if the following conditions hold:

a) None of the expressions E, E_1, E_2, E_3 occurring in for-list elements depend explicitly or implicitly on the controlled variable V.

b) The controlled statement S contains no operations that might change the value of V or of any of the expressions E, E_1, E_2, E_3.

c) None of the for-list elements is a while element.

Under these conditions, which hold for most for-statements occurring in practice, the execution of a for-statement may be described as follows (see, however, 30.3.4 and 30.3.5 below):

30.3.2. *The simplified rule for static for-statements.*

First, an ordered set P of values $R_1, R_2, \ldots, R_m$ (of the same type as the controlled variable) is generated. Every for-list element contributes to this set as follows (the contributions are lined up in the order in which the for-list elements appear):

a) A single expression element «E» contributes the current value of the expression E.

b) A step-until element «E_1 **step** E_2 **until** E_3» contributes the linear sequence running from E_1 with increment E_2 up to at most E_3 (not exceeding this value in the direction of the increment). However, if E_2 and $E_1 - E_3$ have the same sign, the contribution of the step-until element is empty.

Second, the values of the set P are assigned one after the other to the controlled variable V, and for every value of V the controlled statement S is executed once. If P is empty, no execution of S takes place, i.e. the for-statement is then equivalent to a dummy statement.

Third, if during one of the executions of the controlled statement a jump to a destination label outside (or in front of) the for-statement takes place, then the execution of the for-statement is terminated (so-called *termination by a jump*).

Fourth, if the controlled statement has been executed for the last value of the set P, or if P is empty, the execution of the for-statement terminates in a natural way, which we shall refer to as *termination by exhaustion of the for-list.*

30.3.3. *The dynamic rule.* Though not recommended, it is permissible that during execution of a for-statement dynamic effects such as changing the value of the controlled variable occur, e.g.

«**for** $x := 0$ **step** 0.1 **until** 30.05 **do**
  begin
    $\vdots$
    if $y < 10$ **then** $x := x + 0.02$;
    $\vdots$
  end»,

or that the controlled variable enters explicitly into the for-list elements, e.g.
      «**for** $z := 1$ **step** z **until** $200000 + sqrt(z)$ **do** ...».

In all such cases, and in any case where while elements are involved, the precise rule given in 4.6 of the RAR applies. This rule may be pictured as follows (with the same meaning of V, S, E, etc. as in 30.2):

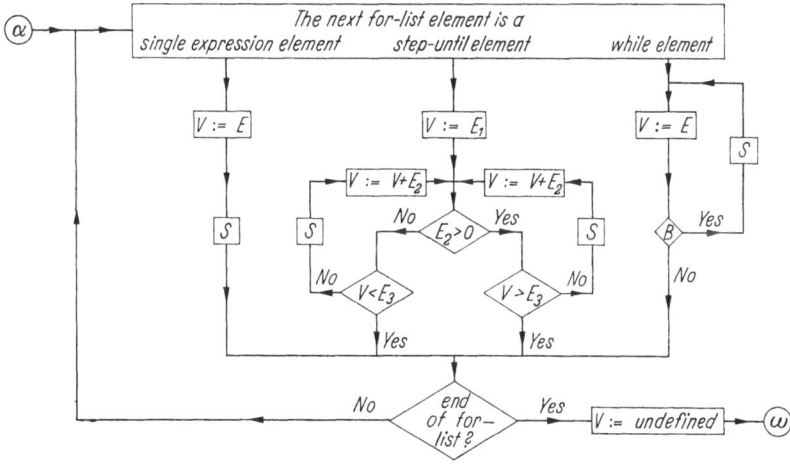

Fig. 27

Upon arrival at ω, the execution of the for-statement is *terminated by exhaustion of the for-list*. Besides this, it can also be *terminated by a jump*, as mentioned in 30.3.2 above.

30.3.4. *Value of the controlled variable after termination.* The following rule applies both to the static as well as to the dynamic case:

a) If the execution of a for-statement terminates by exhaustion of the for-list, the value of the controlled variable is undefined afterwards.

b) If the execution of a for-statement terminates by a jump, then the controlled variable retains the value which it had immediately before the jump, provided the destination of the jump is not outside the scope of the controlled variable (for scope cf. § 42).

These two cases may be exemplified as follows:

«**for** $k := 1$ **step** 1 **until** n **do**
    if $x[k] \neq y[k]$ **then goto** *may* ;
$z[k] := 0$;
   ⋮

   ⋮

may: $z[k] := 1$;
   ⋮
    »

Here the computation continues if the vectors x, y agree completely. The value of k is undefined and the statement $z[k] := 0$ is incorrect.

Continuation if the vectors x, y are different. The value of k is defined and the statement $z[k] := 1$ is correct.

30.3.5. *Jumps into a for-statement.* No jumps from outside into a for-statement are allowed; in other words, the effect of a goto-statement located outside the for-statement, but whose destination is within or in front of the controlled statement, is undefined.

30.4. For-statements and conditional statements

According to the syntax, conditional statements may appear without restrictions as controlled statements of for-statements, but according to 29.2 only the last alternative of an if-else-statement may be a for-statement. Thus

«**for ... do if ... then ... else if ... then ... else ...**»,
«**for ... do if ... then ...**»,
«**if ... then for ... do ...**»,
«**if ... then ... else for ... do ...**»

are all permissible constructions, but

«**if ... then for ... do ... else ...**»

is prohibited by the RAR, namely for the following reasons:

In the original ALGOL-60 report [6] of 1960 there was no such rule restricting the use of for-statements as alternatives of if-else-statements; moreover the report did not state whether the example

$$\text{«if ... then for ... do if ... then ... else ...»} \qquad (1)$$

should be interpreted as an *if-statement* equivalent to

$$\text{«if ... then begin for ... do if ... then ... else ... end»} \qquad (2)$$

or as an *if-else-statement* equivalent to

$$\text{«if ... then begin for ... do if ... then ... end else ...»} \qquad (3)$$

According to the RAR, however, a for-statement cannot be an alternative of an if-else-statement (except the last one), and therefore example (1) now allows only the unique interpretation (2).

30.5. Consequences drawn from the semantic rules

30.5.1. *Influence of roundoff errors.* If a controlled variable is of type **real**, then the computer limitations (cf. 8.2) with regard to real-type variables must be observed. Thus the value of a real-type controlled variable will usually deviate by a small amount from the theoretically expected course, which may have severe and unexpected consequences with respect to the termination of the for-statement. For instance, one would expect that the statement (x declared **real**)

«**for** $x := 0$ **step** 0.01 **until** 0.99 **do** $y := y + 0.01 \times f(x, y)$»,

would cause execution of the controlled statement for $x = 0$, 0.01, 0.02, ..., 0.98, 0.99 and thus integrate — by EULER's method — the differential equation $y' = f(x, y)$ from 0 to 1. However, since the x's are computed numerically and are thus inherently inaccurate, it cannot be predicted — in fact, it depends on the kind of computer to be used — whether in this example

a) instead of $x_{99} = 0.99$ a slightly larger value results and therefore the last execution occurs with $x_{98} = 0.98$ (approximately), or

b) the value x_{99} remains slightly below 0.99; in that case the last execution occurs as expected with the value $x = x_{99}$.

Of course such ambiguities cannot be tolerated in an ALGOL program and must be avoided by programming measures. In the above example one could force an unambiguous decision either by choosing the upper bound between two meshpoints, e.g.

«**for** $x := 0$ **step** 0.01 **until** 0.995 **do** $y := y + 0.01 \times f(x, y)$»,

or by introducing an integer-type controlled variable which counts the number of repetitions (Note that an integer-type controlled variable is stepped precisely):

«**for** $k := 0$ **step** 1 **until** 99 **do** $y := y + 0.01 \times f(k/100, y)$».

Another example where roundoff errors are rather disturbing, is:

«**for** $x := 1.570796326$ **step** $_{10}-12$ **until** 1.570796327 **do** $print(cos(x))$».

Indeed, here the increment is so small that on certain computers it gives no contribution if added to x; as a consequence, a closed loop will result. But even if there is no danger of a closed loop, it is recommended to diminish the influence of roundoff errors by rewriting this statement as

«**for** $k := 0$ **step** 1 **until** 1000 **do** $print(cos(1.570796326 + k/_{10}12))$».

30.5.2. *Empty for-list elements*. In certain circumstances a for-list element is empty and does not cause an execution of the controlled statement. According to the rules given in 30.3 this will occur

  a) if a step-until element is encountered with

$$\text{either} \quad E_1 > E_3 \quad \text{and} \quad E_2 > 0,$$
$$\text{or} \quad E_1 < E_3 \quad \text{and} \quad E_2 < 0.$$

  b) if a while element is taken up in which the first evaluation of the Boolean expression B already yields the value **false**.

If all for-list elements of a for-statement are empty, we speak of an *empty for-statement*. Its for-list is exhausted already before the first execution of S takes place[1]. The execution of such an empty for-statement has no other effect than giving the controlled variable the value "undefined".

30.5.3. *Applications*. The empty for-list element proves to be a very useful concept for many numerical methods. Consider, for instance, the summation $\sum_{\substack{k=1 \\ k \neq j}}^{n} \frac{a_k}{j-k}$. It can be expressed in ALGOL by

«$s := 0$;
  for $k := 1$ **step** 1 **until** $j-1, j+1$ **step** 1 **until** n **do**
$$s := s + a[k]/(j-k)\text{»}.$$

Here k runs from 1 through $j-1$ and then from $j+1$ through n, i.e. from 1 through n with the exception of $k=j$. This is also true for $j=1$ and $k=n$; indeed, for $j=1$ the first for-list element is actually 1 **step** 1 **until** 0 and is therefore empty, while the second for-list element makes the summation from 2 to n.

A further example is the backsubstitution of the Gauss elimination process:

«**for** $k := n$ **step** -1 **until** 1 **do**
  begin
zz: **for** $i := k+1$ **step** 1 **until** n **do** $x[k] := x[k] + a[k, i] \times x[i]$;
    $x[k] := -x[k]/a[k, k]$
  end».

The inner for-statement (statement zz) is empty for the first execution $(k=n)$ of the outer controlled statement; accordingly, the whole example causes execution of the following operations:

---

[1] This means that in ALGOL the jump-out condition of a for-statement is checked at the beginning of a loop. Since this is different in FORTRAN, transcription from ALGOL to FORTRAN must be done with special consideration of possible empty for-list elements.

contributed by

«$x[n] := -x[n]/a[n, n]$; $k = n$

$x[n-1] := x[n-1] + a[n-1, n] \times x[n]$;
$x[n-1] := -x[n-1]/a[n-1, n-1]$;$\left.\rule{0pt}{20pt}\right\}$ $k = n-1$

$x[n-2] := x[n-2] + a[n-2, n-1] \times x[n-1]$;
$x[n-2] := x[n-2] + a[n-2, n] \times x[n]$;$\left.\rule{0pt}{30pt}\right\}$ $k = n-2$
$x[n-2] := -x[n-2]/a[n-2, n-2]$;

$\vdots$ etc.

$x[1] := -x[1]/a[1, 1]$;».

30.5.4. *Jumps inside a for-statement.* While the effect of jumps from inside a controlled statement to the outside and vice versa have been dealt with in 30.3, it remains to discuss the effect of goto-statements whose source and destination are within the same controlled statement. Of course this follows directly from the rules given in 30.3, but will be demonstrated again by the following example:

«**for** $m := -5$ **step** 2 **until** 5 **do**
ra: **begin**
          switch $abcd := ra, rb, rc, rd$;
          $\vdots$
rb: $theta := exp(m)$;
          $\vdots$
          goto $abcd[j]$;
          $\vdots$
rc: **end** m».

For $j = 1$, the statement «**goto** $abcd[j]$» causes a jump to ra with the effect that the controlled statement is re-entered without advancing the value of m and without checking for termination. For $j = 2$, a jump to rb occurs, which produces a little loop inside the controlled statement, of course also without advancing m. For $j = 3$, however, a jump to rc occurs which terminates the present execution of the controlled statement and starts the next execution (with the next value of m), provided the for-list is not yet exhausted.

30.5.5. *Applications of the while element.* Whereas the step-until element serves to repeat the execution of the controlled statement for a strictly linear sequence, the while element was designed to allow for arbitrary stepping of the controlled variable and other dynamic effects. Consider for instance (cf. 31.3):

«**for** $power := 2, 2 \times power$ **while** $power <_{10} 6$ **do begin** ... **end**».

Here the controlled statement is executed exactly 19 times, namely once for every one of the values $power = 2, 4, 8, 16, \ldots, 524288$.

A further example: The *bisection method*.

«**for** $x := (a+b)/2$ **while** $b-a>eps$ **do**
        if $f(x)<0$ **then** $a := x$ **else** $b := x$».

Given an interval $a<x<b$ and a continuous function $f(x)$ with the property $f(a)<0$, $f(b)>0$, this statement computes a new interval (i.e. new values a, b) of length at most eps, such that again $f(a)<0$, $f(b)\geqq0$. The value of eps can be prescribed; however, for too small eps, a closed loop may result unless we extend the jump-out condition of the while element to «$b-a>eps \wedge x \neq a \wedge x \neq b$».

Thus, undoubtedly, the while element allows a more condensed and elegant description of certain computing processes. However, this must be paid for by a loss of clarity, and therefore a too extensive use of this instrument is not recommended.

30.6. Efficiency considerations

For-statements contribute heavily to the total computing time of ALGOL programs. This is especially true if for-statements are nested; in such cases, obviously the innermost loops (e.g. controlled statement 2 in example 30.1.3) contribute heaviest. As a consequence the programmer should keep time-consuming operations out of innermost loops whenever possible. Often this cannot be achieved, but sometimes a rearrangement of the running subscripts may help.

30.6.1. In the summation process

«$s := 0$;
  for $i := 1$ **step** 1 **until** n **do**
    for $j := 1$ **step** 1 **until** i **do**
      for $k := j$ **step** 1 **until** i **do**
        $s := s+f(i,j) \times g(j,k)$ »

the computing time is roughly on the order of $n^3/6$ times the evaluation time for the two function designators $f(i,j)$ and $g(j,k)$. Obviously this piece of program becomes more efficient if we take the function designator $f(i,j)$ out of the innermost loop:

«$s := 0$;
  for $i := 1$ **step** 1 **until** n **do**
    for $j := 1$ **step** 1 **until** i **do**
    begin
      $fij := f(i,j)$;
      for $k := j$ **step** 1 **until** i **do** $s := s+fij \times g(j,k)$
    end i *and* j».

If the evaluation time is much greater for g than for f, this modification offers not much of a saving; in this case, it would be better to rearrange the hierarchy of the subscripts i, j, k: We note that they must meet the condition $j \leq k \leq i$ but otherwise may run arbitrarily from 1 to n. Hence

```
«s := 0 ;
  for j := 1 step 1 until n do
    for k := j step 1 until n do
    begin
      gjk := g(j, k) ;
      for i := k step 1 until n do s := s + gjk × f(i, j)
    end j and k»
```

is a valid rearrangement which reduces the computing time to about $n^3/6$ evaluations of f plus $n^2/2$ evaluations of g.

30.6.2. Similarly, where conditional elements appear in a controlled statement, it should be attempted to delegate as much of the checking as possible to the outer loops. As an example,

```
«for i := 2 step 1 until n do
  for j := 1 step 1 until i − 1 do
    if a[i − j] > 0 ∧ a[j] < 0 then c[i, j] := false»
```

can be rewritten less elegantly but in general more efficiently as

```
«for j := 1 step 1 until n − 1 do
  if a[j] < 0 then
    for i := j + 1 step 1 until n do
      if a[i − j] > 0 then c[i, j] := false»,
```

or, by introducing a new variable $k = i - j$:

```
«for k := 1 step 1 until n − 1 do
  if a[k] > 0 then
    for j := 1 step 1 until n − k do
      if a[j] < 0 then c[k + j, j] := false».
```

Of course, the saving achieved in the above example depends heavily upon the relative frequencies of the fulfilled conditions, $a[k] > 0$ and $a[j] < 0$, which are encountered during the process.

30.6.3. Quite often the efficiency of loops can be improved just by choosing other subscripts, i.e. by performing an affine transformation in the "subscript space". Consider, for instance, the statement

```
«for i : = 1 step 1 until n do
  for j := i + 1 step 1 until n do p[j − i] := a[j − i, i]»,
```

which, by introducing a new variable $k = j - i$, can be rewritten as

«**for** $i := 1$ **step** 1 **until** n **do**
   for $k := 1$ **step** 1 **until** $n - i$ **do** $p[k] := a[k, i]$».

In doing so we have eliminated the evaluation of the subscript expressions $j - i$ in the inner loop. Of course this gives only a very slight improvement here, but it may serve to indicate what might be done in less trivial situations.

30.6.4. Let us consider example 30.1.4, which computes the components $c[i]$ of the product matrix × vector *serially*. It would seem that we could just as well compute these components also in *parallel*, namely by

«**for** $j := 1$ **step** 1 **until** n **do** $c[j] := 0$;
   for $j := 1$ **step** 1 **until** n **do**
      for $i := 1$ **step** 1 **until** n **do** $c[i] := c[i] + a[i, j] \times b[j]$».

Indeed, the number and kinds of operations involved are the same in both cases; all the same, the version given in 30.1.4 is preferable for the following reason: If one would like to economize the summation over j with a code procedure (cf. 47.4), this can be achieved only if this summation is carried out as an unbroken process:

«$c[i] := 0$;
   for $j := 1$ **step** 1 **until** n **do**   (4)
      $c[i] := c[i] + a[i, j] \times b[j]$».

Indeed, an important part of the economisation is the elimination of the repeated storage reference to $c[i]$, which causes a value to be put into storage and then read again from storage immediately afterwards. In an optimized code procedure for performing the task (4), the partial sums would be kept in the accumulator (which would not be possible for parallel summation), and only the final sum would be stored as $c[i]$.

30.6.5. Finally, we observe that the statement

«**for** $j := 1$ **step** 1 **until** *upper* **do** ...»,

where *upper* is a function procedure without formal operands, is somewhat uneconomical. Indeed, the test for termination, hence the evaluation of the function designator «*upper*», is performed for every j. If the order of j is irrelevant, we can write

«**for** $j := upper$ **step** -1 **until** 1 **do** ...»

instead, in which case «*upper*» is evaluated only once.

Chapter V

Miscellaneous Applications

We have now collected sufficient material so that we can give somewhat more complicated examples selected from various fields of applications. However, since declarations still have not been treated, these examples cannot be presented as complete ALGOL programs but rather as
program fragments, beginning at a point where all declarations have
been given and all input operations have been performed, and ending
at a point where the results are ready for output. Furthermore, these
examples, though correct in principle, are not sufficiently foolproof for
actual use. In the following this will be understood without further
mentioning.

§ 31. Algebraic Problems

31.1. Gauss elimination

The elementary process of Gauss[1] for solving $A\,\vec{x}+\vec{b}=0$ consists of
3 parts, namely 1) the *elimination proper*, which means splitting the
matrix A into two triangular factors B, C; 2) the *forward substitution*,
i.e. the operation $\vec{r} := C^{-1}\vec{b}$, and 3) the *backsubstitution*, i.e. the operation
$\vec{x} := -B^{-1}\vec{r}$.

It has been found convenient to organize the elimination in such a
way that the matrices B and C are stored together as the BC-matrix
in the same place as A; the latter is therefore overwritten by the process.
Accordingly, in our program the $a[i, k]$, while initially being elements
of A, are later elements of B or C, depending upon whether $i \leq k$ or $i > k$.
Similarly, all 3 vectors $\vec{b}, \vec{r}, \vec{x}$ are stored as one and the same array
$s[1:n]$: at the beginning the $s[i]$ must be given as the constant terms
$b_1, \ldots, b_n$ of the system; after the elimination they are the vector $\vec{r}$, and
at the very end the $s[1], s[2], \ldots, s[n]$ are the solution $x_1, x_2, \ldots, x_n$.

If step 2) is incorporated into step 1), the following program fragment
(which, however, does not search for pivots) is obtained:

«**begin**
    for $j := 1$ **step** 1 **until** $n-1$ **do**
        for $i := j+1$ **step** 1 **until** n **do**

---

[1] See § 6 in [*44*].

```
elim:    begin
c1:          a[i, j] := − a[i, j]/a[j, j] ;
             for k := j+1 step 1 until n do
                 a[i, k] := a[i, k]+a[i, j]×a[j, k] ;
             s[i] := s[i]+a[i, j]×s[j]
         end elim ;

back:
         for k := n step − 1 until 1 do
         begin
zz:          for i := k+1 step 1 until n do
                 s[k] := s[k]+a[k, i]×s[i] ;
c2:          s[k] := − s[k]/a[k, k]
         end back ;
end».
```

Elimination of x_j from i-th equation by adding appropriate multiple of j-th equation to the i-th.

Computes $x_k = s[k]$ from the k-th equation of the triangular system $B\vec{x} + \vec{r} = 0$.

In this elimination scheme all $a[i, k]$ with $i, k > j$ are modified in the j-th elimination step by adding the product $a[i, j] \times a[j, k]$. In contrast to this, T. BANACHIEWICZ (cf. [44], § 6.2) postpones all operations upon $a[i, k]$ until $j = min(i, k) − 1$, but then adds all products at once to $a[i, k]$, i.e. performs the operation

$$a[i, k] := a[i, k] + a[i, 1] \times a[1, k] + a[i, 2] \times a[2, k] + \cdots$$
$$+ a[i, l] \times a[l, k],$$

where $l = min(i, k) − 1$. Moreover, if $k < i$, $a[i, k]$ is divided by $− a[k, k]$ (in the above program this operation would be performed in the next elimination step by statement $c1$).

The Banachiewicz scheme requires a complete change in the hierarchy of the i-, j- and k-loops: First we let i, k run through all matrix positions, and for every i, k all operations upon $a[i, k]$ are performed (the constant terms are treated similarly):

```
«begin
    for i := 2 step 1 until n do
    begin
        for k := 1 step 1 until n do
        begin
            l := if i > k then k−1 else i−1 ;
            t := a[i, k] ;
sum:        for j := 1 step 1 until l do
                t := t+a[i, j]×a[j, k] ;
            a[i, k] := if k < i then − t/a[k, k] else t ;
        end k ;
```

operations upon $a[i, k]$.

```
    for j := 1 step 1 until i − 1 do          ⎫ operations
        s[i] := s[i] + a[i, j] × s[j] ;       ⎬ upon s[i].
```
    end i ;
comment Backsubstitution is the same as before ; ».

The Banachiewicz scheme has the advantage that it allows economizing the inner product loop (statement *sum*) with the aid of code procedures (cf. 47.4); on the other hand, pivot strategy is much more complicated than for the original Gauss scheme.

31.2. Newton's method for algebraic equations

For given values of the coefficients $a[0], a[1], \ldots, a[n]$ of a polynomial $f(x) = \sum_{k=0}^{n} a[k] x^k$ and with a given initial value x (preferably close to a root) the following piece of program attempts to compute a root:

```
«begin
rep:  f := g := 0 ;
horn: for k := n step −1 until 0 do
      begin
          g := g × x + f ;
          f := f × x + a[k]
      end ;
      delta := − f/g ;
      x := x + delta ;
      if abs (delta) > eps then goto rep
end».
```

The for-statement labelled *horn* computes the value f and the derivative g of $f(x)$ at x; indeed, for each k we have (after termination of the controlled statement)

$$f = \sum_{j=k}^{n} a[j] \, x^{j-k}, \qquad g = \sum_{j=k}^{n} (j-k) \, a[j] \, x^{j-k-1},$$

as may be verified by induction.

We note in passing that the termination criterion of the above piece of program by no means meets the requirements of computing practice because it is usually impossible to give an a priori value *eps* such that the jump back to *rep* occurs just as long as this is both necessary and meaningful. Indeed, a too small *eps* tends to cause a closed loop with x jumping around the root in an erratic manner, whereas a too big *eps* discontinues the process while it is still capable of improving the approximation of x. In § 36 we shall develop a program which does not exhibit this kind of behavior.

31.3. The Dandelin-Graeffe method

The basic idea of this method is to transform a given polynomial $f(x)$ into one whose roots are the squares of the former and to repeat this process until a polynomial is obtained, the roots of which are so strongly separated that they are practically the quotients of consecutive coefficients (e.g. the roots of $x^4 - 10^{50} x^3 + 10^{80} x^2 - 10^{70} x + 10^{20} = 0$ are approximately 10^{50}, 10^{30}, 10^{-10}, 10^{-50}).

The roots r_i of the original equation are then the 2^k-th roots of the zeros of the last polynomial if k root-squaring steps have been needed to obtain it. Complex roots are more difficult to compute with this method; only their moduli are obtained easily. One root-squaring step, i.e. the step from

$$\sum_k a_k (-x)^k = a_n \prod_i (x - r_i) \quad \text{to} \quad \sum_k b_k (-x)^k = b_n \prod_i (x - r_i^2)$$

is, as an example, described for $n=6$ by the following formulae

$$
\begin{aligned}
b_0 &= a_0^2 \\
b_1 &= a_1^2 - 2 a_0 a_2 \\
b_2 &= a_2^2 - 2 a_1 a_3 + 2 a_0 a_4 \\
b_3 &= a_3^2 - 2 a_2 a_4 + 2 a_1 a_5 - 2 a_0 a_6 \\
b_4 &= a_4^2 - 2 a_3 a_5 + 2 a_2 a_6 \\
b_5 &= a_5^2 - 2 a_4 a_6 \\
b_6 &= a_6^2 .
\end{aligned}
\tag{1}
$$

Afterwards the b_j are again denoted by a_j and the process is repeated. It can be stopped as soon as in the computation of b_j the other terms become negligible with respect to a_j^2, and this for all j. If this happens for only one j, this is recorded by setting the j-th component of a Boolean vector sep to **true** and means that we could split the equations into one with roots $r_1, r_2, \ldots, r_j$ and one with $r_{j+1}, r_{j+2}, \ldots, r_n$. If $sep[j-1]$ and $sep[j]$ are both **true**, we can compute the modulus of one root r_j and reduce the order of the equations by one. These measures are included in the following program, but it does not contain measures against the very small and very large numbers which usually occur with this method and threaten to discontinue the process by overflow of the exponent[1].

«**for** $power := 2, 2 \times power$ **while** $power < {}_{10}6$ **do**
begin
  comment follows one root-squaring step as indicated by formulae (1);
    for $j := 0$ **step** 1 **until** n **do**

---

[1] A Graeffe-like method which avoids the occurrence of very small and very large numbers has been described by GRAU [16].

begin
    $b1 := b[j] := a[j]\uparrow2$;
    $sep[j] :=$ **true** ;
    $s := -2$;
    for $i := 1$ **step** 1 **until** (**if** $j > n-j$ **then** $n-j$ **else** j) **do**
    begin
        $b[j] := b[j] + s \times a[j-i] \times a[j+i]$;
        $s := -s$;
        $sep[j] := sep[j] \wedge (b1 = b[j])$
    end i
  end j ;
  for $k := 0$ **step** 1 **until** n **do** $a[k] := b[k]$;
comment follows reduction of equation, if consecutive sep's are **true**.
        i counts eliminated coefficients ;
  $i := 0$;
  $quot := 1$;
reduce:
  for $k := 1$ **step** 1 **until** n **do**
    if $sep[k-1] \wedge sep[k]$ **then**
    begin
        $mod[n-i] := a[k-1]/a[k]$;
        $quot := quot \times mod[n-i]$;
        $mod[n-i] := exp\,(ln\,(mod[n-i])/power)$;
        $i := i+1$
    end
    else
        if $i \neq 0$ **then** $a[k-i] := quot \times a[k]$;
  $n := n-i$
end *power* ».

We observe that statement *reduce* would transform the situation

$$n = 6, \quad p = 32, \quad a[0:n] = (1, {}_{10}50, {}_{10}60, {}_{10}70, {}_{10}60, {}_{10}25, 1),$$
$$sep[0:n] = (\textbf{true, true, false, true, true, false, true})$$

into

$$n = 4, \quad a[0:n] = (1, {}_{10}10, {}_{10}20, {}_{10}{-}15, {}_{10}{-}40)$$

and produces the moduli of two roots:

$$mod[6] = 10^{-1.5625}, \quad mod[5] = 10^{0.3125}.$$

31.4. The stability criterion of Routh

The following compound statement decides, for given coefficients $a[k]$ of a polynomial $\sum_{0}^{n} a_k x^k$, whether all its roots have negative real parts (in which case it produces the logical value *stable* = **true**) or not (in which case *stable* = **false**).

If the computation is done by hand, it is recommended (viz. ZUR-MUEHL [43], p. 82/83) to arrange the principal minors of the Hurwitz determinant in a Routh-table (e.g. for even n):

$$
\begin{array}{llllll}
a_0 & a_2 & a_4 & a_6 & \ldots\ldots & a_{n-2} \; a_n \\
a_1 & a_3 & a_5 & a_7 & \ldots\ldots & a_{n-1} \; 0 \\
b_2 & b_4 & b_6 & b_8 & \ldots\ldots & b_n \quad\; 0 \\
c_3 & c_5 & c_7 & c_9 & \ldots c_{n-1} & 0 \quad\;\; 0 \\
d_4 & d_6 & d_8 & d_{10} & \ldots d_n & 0 \quad\;\; 0
\end{array}
$$
$$\text{etc.}$$

The present program reflects the staircase shape of the Routh-table, but all quantities $a_k, b_k, c_k, d_k \ldots$ for the same k are stored as the same component $a[k]$ of an array a. This is possible since e.g. b_6 is no longer needed as soon as d_6 has been computed.

```
«begin
    stable := false ;
    for j := 0 step 1 until n − 1 do
    begin
        if a[0] × a[j + 1] ≤ 0 then goto ex ;
        c := − a[j]/a[j + 1] ;
        for k := j + 2 step 2 until n − 1 do a[k] := a[k] + c × a[k+1]
    end ;
    stable := true ;
ex:
  end».
```

§ 32. Interpolation and Numerical Quadrature

32.1. Neville-Lagrange interpolation

Let a polynomial of degree n be defined by two vectors $a[0:n]$, $b[0:n]$ representing the coordinates of $n + 1$ points on the curve $y = f(x)$. We have many methods to compute the value of f at x, one of them being NEVILLE's scheme [24], which is based on a relation between all polynomials $f_{ij}(x)$, where $f_{ij}(x)$ is of order $j - i$ and defined by $f_{ij}(a[k]) = b[k]$ for $k = i, i+1, \ldots, j-1, j$.

Indeed, if we introduce the values $y_{i,j} = f_{ij}(x)$ $(i, j = 1, 2, \ldots, n; \; i \leq j)$, then

a) $y_{k,k} = b[k]$ $\quad (k = 0, 1, \ldots, n)$,

b) $y_{i,j} = y_{i+1,j} + \dfrac{x - a[j]}{a[j] - a[i]} (y_{i+1,j} - y_{i,j-1})$ $\quad$ (for all i, j; $i \leq j$),

c) $f = y_{0,n}$ is the required value $f(x)$.

In our program $y_{i,j}$ will be denoted by $y[j]$ since $y_{i-1,j}$ is no longer needed after $y_{i,j}$ has been computed.

```
«begin
    for j := 0 step 1 until n do y[j] := b[j] ;
aa: for k := 1 step 1 until n do
bb:     for j := n step −1 until k do
cc:         y[j] := y[j]+(x−a[j])×(y[j]−y[j−1])/(a[j]−a[j−k]) ;
    f := y[n]
  end».
```

If the values $y_{i,j}$ are arranged in a *Neville-table*

$$
\begin{array}{llll}
y_{0,0} & & & \\
 & y_{0,1} & & \\
y_{1,1} & & y_{0,2} & \\
 & y_{1,2} & & y_{0,3} \\
y_{2,2} & & y_{1,3} & \quad \cdot \\
 & y_{2,3} & & y_{1,4} \quad \vdots \quad \cdot \quad \vdots \; y_{0,n}=f \\
y_{3,3} \quad \vdots & & y_{2,4} \quad \vdots & \quad \cdot \\
\vdots & & \vdots & \\
 & & & \\
y_{n,n} & & &
\end{array}
$$

then every execution of the for-statement bb (controlled statement of aa) causes generation of a new column of this table, namely the one containing the values $y[j]=y_{j-k,j}$ $(j=k, k+1, \ldots, n)$. At the end, the array $y[0:n]$ contains the top row of the Neville table, and in particular $y[n]$ is the required value $f(x)$.

Another method to interpolate with the same given data is the *barycentric formula*[1], for which we obtain the following program:

```
«begin
weights:
    for j := 0 step 1 until n do
    begin
  comment prepare weight w[j]=1/product (over k ≠ j) of (a[j]− a[k]) ;
        w[j] := 1 ;
        for k := 0 step 1 until j−1, j+1 step 1 until n do
                w[j] := w[j]/(a[j]− a[k])
    end j ;
evaluate:
    s := t := 0 ;
    for k := 0 step 1 until n do
```

---

[1] W. J. TAYLOR [37].

begin
comment add one new term to each numerator and denominator of
        barycentric formula ;
    $d := x - a[k]$;
    if $d = 0$ **then** $d := {}_{10}-30$;
    $s := s + b[k] \times w[k]/d$;
    $t := t + w[k]/d$
end k ;
  $f := s/t$
end ».

This program, though considerably longer than that for NEVILLE's
method, is more economical if the same polynomial must be evaluated
frequently or if different polynomials, given at the same abscissae $a[k]$,
must be interpolated. Indeed, for-statement *weights*, which is the only
part which requires a computing time on the order of $O(n^2)$, depends
neither upon the $b[k]$ nor upon x and therefore can be executed once
and for all as long as the $a[k]$ do not change. All later interpolations for
the same a's can be done by a jump to *evaluate* and require only a
computing time of order $O(n)$. This compares favorably to NEVILLE's
method, which always requires a computing time on the order of $O(n^2)$.

We note in passing that for large n the multiplication of the many
small differences $a[j] - a[k]$ may cause an underflow of the exponent;
this may require special countermeasures.

32.2. Hermite interpolation with equidistant abscissae

The virtues of Hermite interpolation (function given not only by
values but also by first derivatives at the mesh-points) are too well
known to require further discussion. Let y_k and y_k' be the given values
of $f(x)$ and $f'(x)$ at $x_k = x0 + k \times h$ ($k = 0, 1, \ldots, n$). Then the value of
the Hermite interpolation polynomial $H(x)$ of order $2n+1$ at x can be
expressed in our specialisation by the following barycentric type of
formula[1]:

$$H(x) = \frac{\sum_{k=0}^{n}\left\{\frac{A_k}{(z-k)^2} + \frac{B_k}{(z-k)}\right\}f_k + h\sum_{k=0}^{n}\frac{A_k}{(z-k)}f_k'}{\sum_{k=0}^{n}\left\{\frac{A_k}{(z-k)^2} + \frac{B_k}{(z-k)}\right\}} \tag{1}$$

where

$$A_k = \left(\frac{n}{k}\right)^2, \quad z = (x - x_0)/h,$$

$$B_k = 2A_k(\varphi_{n-k} - \varphi_k),$$

---

[1] See KUNTZMANN [23], p. 169ff.

with

$$\varphi_k = 1 + \frac{1}{2} + \frac{1}{3} \cdots + \frac{1}{k}.$$

This leads to the following program ($yl[k]$ representing y'_k):

```
«begin
    phi[0] := 0 ;
    for k := 1 step 1 until n do phi[k] := phi[k−1]+1/k ;
    s := t := 0 ;
    w := 1 ;
    z := (x − x0)/h ;
    for k := 0 step 1 until n do
    begin
        if z − k = 0 then begin bigh := y[k] ; goto ex end ;
comment add one further term to both numerator s and denominator t
        of barycentric formula (1) ;
        r := 1/(z − k)↑2 − 2×(phi[k] − phi[n − k])/(z − k) ;
        s := s + w×(r×y[k]+h×yl[k]/(z − k)) ;
        t := t + w×r ;
        w := w×(n − k)↑2/(k + 1)↑2 ;
    end k ;
    bigh := s/t ;
ex:
end».
```

32.3. Newton interpolation in an equidistant table

Let $a[0:bign]$ be an array representing an extended table of a function $f(x)$, such that $a[k]$ is the value of f at $x_k = x0 + k \times h$, where $x0$ and h are also given. It would be uneconomical and numerically unstable to evaluate the full interpolation polynomial of order $bign$ for a given x. Instead we compute $f(x)$ by Newton-Gregory interpolation from the values of f at eight of the abscissae x_j, four on either side of the given x. This implies 1) selection of proper abscissae x_k through x_{k+7}, i.e. computation of k (statement sel), 2) building the difference table for the values $a[k]$ through $a[k+7]$ (statement dif), and 3) evaluation of the Newton-Gregory formula (statement eva):

```
«begin
    t := (x − x0)/h ;
sel: k := if t < 3.5 then 0 else if t > bign − 3.5 then
            bign − 7 else entier (t) − 3 ;
    t := t − k ;
    for i := 0 step 1 until 7 do y[i] := a[k+i] ;
```

dif: **for** $i := 1$ **step** 1 **until** 7 **do**
      for $j := 7$ **step** -1 **until** i **do** $y[j] := y[j] - y[j-1]$;
  $f := y[7]$;
eva: **for** $i := 6$ **step** -1 **until** 0 **do** $f := f \times (t-i)/(i+1) + y[i]$
end ».

32.4. Romberg Quadrature

The Romberg method for computing $\int_a^b f(x)\,dx$ has become known through a number of recent papers[1]. Its main feature is the T-table

$$
\begin{array}{cccc}
T_0^{(0)} & & & \\
 & T_1^{(0)} & & \\
T_0^{(1)} & & T_2^{(0)} & \\
 & T_1^{(1)} & & T_3^{(0)} \\
T_0^{(2)} \quad \vdots & & T_2^{(1)} \quad \vdots & \\
\vdots & & \vdots &
\end{array}
$$

which in fact is a Neville scheme (for $x=0$) of a function $T(x)$ given by the values $T(4^{-k}) = T_0^{(k)}$ (= trapezoidal values for subdivision of the quadrature interval into 2^k equal parts). The Neville formula reduces in this case to

$$
T_m^{(k)} = \frac{4^m\, T_{m-1}^{(k+1)} - T_{m-1}^{(k)}}{4^m - 1},
$$

which together with the evaluation of the $T_0^{(k)}$ yields the following program (it is assumed that $f(x)$ is a function designator which produces the value of the integrand f at x):

```
«begin
    n := 1 ;
    t[0] := (b − a)/2 × (f(a) + f(b)) ;
    for k := 1 step 1 until m do
    begin
        n := 2×n ;
        h := (b − a)/n ;
        p := 4 ;
        s := 0 ;
accu:   for i := 1 step 2 until n do
                s := s + f(a + i×h) ;
        t[k] := t[k − 1]/2 + s×h ;
```

} Evaluation of trapezoidal rule.

---

[1] See for instance BAUER, RUTISHAUSER, STIEFEL [8].

for $j := k - 1$ **step** -1 **until** 0 **do**
  begin
    $t[j] := (p \times t[j+1] - t[j])/(p - 1)$;
    $p := 4 \times p$
  end j

} Romberg step for computing new anti-diagonal of T-table.

  end k ;
  $int := t[0]$
end »

The structure of this program is slightly different from the Neville program given in 32.1. Indeed, here it is natural to compute the $T_0^{(0)}, T_0^{(1)}, T_0^{(2)}, \ldots$ etc. in this order and then to compute immediately after every $T_0^{(k)}$ the values $T_1^{(k-1)}, T_2^{(k-2)}, \ldots, T_k^{(0)}$, where $T_{k-j}^{(j)}$ appears in our program as $t[j]$. With the ALGOL notations the T-table will therefore appear as follows:

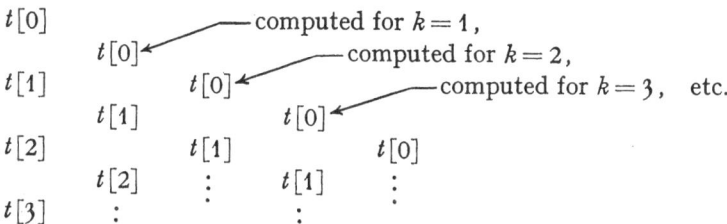

Unfortunately this program has — like most other programs for ROMBERG's method published heretofore — very poor properties with respect to accumulation of roundoff errors. To improve this situation, we could replace the for-statement *accu*, which is responsible for this imperfection, with the following compound statement, which avoids too frequent additions of small terms to a large partial sum. Indeed, statement *f0* accumulates at most 16 terms of the trapezoidal sum, while statement *f1* collects the contributions of at most 16 such presummations, and finally statement *f2* accumulates all terms produced by statement *f1*.

«**begin**
  $n0 :=$ **if** $n > 32$ **then** 32 **else** n ;
  $n1 :=$ **if** $n > 512$ **then** 512 **else** n ;
f2: **for** $k2 := 1$ **step** 512 **until** n **do**
    begin
      $s1 := 0$;
f1:   **for** $k1 := k2$ **step** 32 **until** $k2 + n1 - 1$ **do**
      begin
        $s0 := 0$;

```
f0:         for k0 := k1 step 2 until k1+n0−1 do
                s0 := s0+f(a+h×k0) ;
            s1 := s0+s1
        end k1 ;
        s := s+s1
    end k2 ;
end».
```

§ 33. Numerical Integration of Differential Equations

33.1. Runge-Kutta method, Nystroem modification

An advantageous method for integrating ordinary differential equations numerically is the Runge-Kutta method [17]. The following piece of program corresponds to a modification given by NYSTROEM [25] which integrates directly differential equations of second order:

Let $y''=f(x, y, y')$ be the differential equation, $x, y, y1$ the given initial values $x, y(x)$ and $y'(x)$, h the length of the integration step and p the number of such steps to be performed. It is assumed that $f(x, y, y1)$ is a function designator which produces the value of the second derivative ($y2$ in our program) for given argument x, function value y and derivative $y1$.

Like the classic Runge-Kutta method, the Nystroem algorithm also uses three auxiliary points A, B, C within the integration step from x_k to x_{k+1}. The values y, y', y'' at those places are denoted by ya, yb, yc, $ya1, yb1, yc1, ya2, yb2, yc2$, whereas the corresponding values at the meshpoints proper are denoted by $y, y1, y2$.

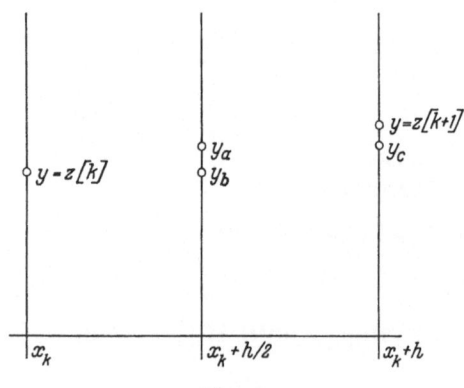

Fig. 28

«**for** $k := 1$ **step** 1 **until** p **do**
  begin
    $y2 := f(x, y, y1)$;
      $x := x + h/2$;
    $ya := y + h \times y1/2 + h{\uparrow}2 \times y2/8$;$\left.\vphantom{\begin{matrix}a\\a\\a\end{matrix}}\right\}$ Auxiliary point A.
    $ya1 := y1 + h \times y2/2$;
    $ya2 := f(x, ya, ya1)$;
    $yb := ya$;$\left.\vphantom{\begin{matrix}a\\a\\a\end{matrix}}\right\}$ Auxiliary point B.
    $yb1 := y1 + h \times ya2/2$;
    $yb2 := f(x, yb, yb1)$;
      $x := x + h/2$;
    $yc := y + h \times y1 + h{\uparrow}2 \times yb2/2$;$\left.\vphantom{\begin{matrix}a\\a\\a\end{matrix}}\right\}$ Auxiliary point C.
    $yc1 := y1 + h \times yb2$;
    $yc2 := f(x, yc, yc1)$;
    $z[k] := y := y + h \times y1 + h{\uparrow}2 \times (y2 + ya2 + yb2)/6$;$\left.\vphantom{\begin{matrix}a\\a\\a\end{matrix}}\right\}$ Completion of integration step.
    $z1[k] := y1 := y1 +$
        $+ h \times (y2 + 2 \times ya2 + 2 \times yb2 + yc2)/6$;
  end».

After termination the values of y, y' at all meshpoints $x + k \times h$ ($k = 1, 2, \ldots, p$) are available as components of the arrays $z[1:p]$ and $z1[1:p]$.

33.2. The Adams-Bashforth method

The problem is to integrate $y' = f(x, y)$ by the Adams extrapolation method of order q (open $q+1$ point formula) over r steps of length h, whereby the initial values $x_0, y(x_0) = y_0$ are given. First we must integrate over $q-1$ steps by some other method (e.g. RUNGE-KUTTA) in order to obtain the necessary starting values $x_{q-1}, y_{q-1}, y'_{q-1}, y'_{q-2}, \ldots, y'_1, y'_0$ (in fact, we shall assume that these values are given at the beginning as $x, y, z[q-1], z[q-2], \ldots, z[1], z[0]$). Then we can continue with the recursion formula

$$y_k = y_{k-1} + h \times (c_0 y'_{k-1} + c_1 y'_{k-2} + \cdots c_{q-1} y'_{k-q}) \quad (k = q, q+1, \ldots, r). \quad (1)$$

In this formula the c's are fixed for given q and can be generated prior to the integration with the aid of the generating function

$$\sum_{k=0}^{\infty} b_k t^k = -t/((1-t) \times ln(1-t)),$$

and

$$\sum_{k=0}^{q-1} b_k (1-t)^k = \sum_{k=0}^{q-1} c_k t^k.$$

The recursion formula (1) is easily programmed for a computer; however, the storage of the derivatives y_i' $(i = q, q+1, \ldots, r)$, which are produced by the process and for which at any time the last q must be available, presents a problem:

If the y_{k-j}' are stored as $z[q-j]$, as (1) would suggest, we would have to shift them by one storage place whenever k is increased; this saves storage but wastes time. If the y_i' are stored as $z[i]$ for all i, we need not shift them but hereby waste storage space in a way which becomes prohibitive if a system of differential equations is to be integrated by the same method. We therefore decide to store the y_{k-j}' cyclically, namely y_{k-j}' as $z[i]$, where $0 \le i < q$ and $i \equiv k-j \pmod{q}$. There is a little inconvenience with this scheme insofar as the summation loop for evaluating (1) is split into two loops, but even this can be avoided at the expense of storing the coefficients c_i twice, namely c_i as $c[i]$ and at the same time as $c[i+q]$ $(i = 0, 1, \ldots, q-1)$. This is shown by the following confrontation of ordinary versus ALGOL notation for the factors appearing in the terms of (1) (Note that $p \equiv k-1 \pmod{q}$):

$$
\begin{matrix}
\text{ordi-} \\ \text{nary}
\end{matrix}
\left\{
\begin{matrix}
c_p & c_{p-1} & \cdots c_1 & c_0 & \\
y_{k-p-1}' & y_{k-p}' & \cdots y_{k-2}' & y_{k-1}' &
\end{matrix}
\right.
\left|
\begin{matrix}
c_{q-1} & \cdots c_{p+2} & c_{p+1} \\
y_{k-q}' & \cdots y_{k-p-3}' & y_{k-p-2}'
\end{matrix}
\right.
$$

$$
\text{ALGOL}
\left\{
\begin{matrix}
c[p+q] & c[p+q-1] & \cdots c[q+1] & c[q] & \\
z[0] & z[1] & \cdots z[p-1] & z[p] &
\end{matrix}
\right.
\left|
\begin{matrix}
c[q-1] & . & c[p+2] & c[p+1] \\
z[p+1] & . & z[q-2] & z[q-1]
\end{matrix}
\right.
$$

Applying these ideas to a system of differential equations, which we assume to be defined by a procedure *equ* such that a call «*equ*(x, y, n) *res*: (f)» produces the derivatives $z[k] = \dfrac{d}{dx} y[k]$ (cf. 44.7.3), we obtain the following piece of program:

```
«begin
    p := q − 1 ;
    for k := q step 1 until r do
    begin
comment here p congruent k − 1 (mod q) ;
        for l := 1 step 1 until n do
        begin
comment Evaluation of Adams-Bashforth formula ;
            s := 0 ;
            for j := 0 step 1 until q − 1 do s := s + c[p+q−j] × z[j,l] ;
            yy[k, l] := y[l] := y[l] + h × s ;
        end l ;
        x := x + h ;
        p := if p = q − 1 then 0 else p + 1 ;
```

comment Compute derivatives $f[l]$ of $y[l]$ at x and insert these into
      the array z ;
      $equ\ (x, y, n)\ res: (f)$;
   for $l:=1$ **step** 1 **until** n **do** $z[p, l]:=f[l]$
  end k
end ».

After termination, the **array** $yy[q:r, 1:n]$ contains the components of
the solution at all meshpoints $x_q, x_{q+1}, \ldots, x_r$.

33.3. Laplace's equation

Let L be a domain whose boundary lies entirely on grid lines of a
square grid (with mesh size $h \times h$). We attempt to solve

$$\Delta u = f(x, y) \quad \text{in } L,$$
$$u = 0 \qquad \text{on boundary}.$$

Such a domain can be described in terms of the intersections of all
vertical grid-lines with the domain. Indeed, if we define the interior grid
points P_{ij} for every fixed i by the inequalities

$$a_1 \leqq j \leqq a_2$$
$$a_3 \leqq j \leqq a_4$$
$$\vdots$$
$$a_{m-1} \leqq j \leqq a_m \quad (m \text{ even}),$$

and represent the a_p for every i as a vector $\vec{v}_i = (a_1, a_2, \ldots, a_m)$, the
domain is defined. For convenience the number m is added as the
zero-th component a_0 to this vector.

However, this "description" of a domain usually takes an undue
amount of storage space, which can be reduced considerably if identical
vectors $\vec{v}_i$ are listed only once, but a $-$2-nd and a $-$1-st component
are added with the meaning that $\vec{v}_i$ is valid for those i for which
$a_{-2} \leqq i \leqq a_{-1}$. The whole domain is then defined by an **integer array**
$a[1:n, -2:mmax]$, every line of which pertains to a group of vertical
grid lines which have identical intersections with the interior of L.

As an example the domain (see Fig. 29) is defined by the array
($n=6$, $mmax=6$)

	-2	-1	0	1	2	3	4	5	6
1	5	6	2	5	17	0	0	0	0
2	7	12	4	5	17	39	45	0	0
3	13	13	4	5	17	32	45	0	0
4	14	18	6	5	9	15	17	32	45
5	19	28	4	5	9	15	45	0	0
6	29	33	2	5	45	0	0	0	0

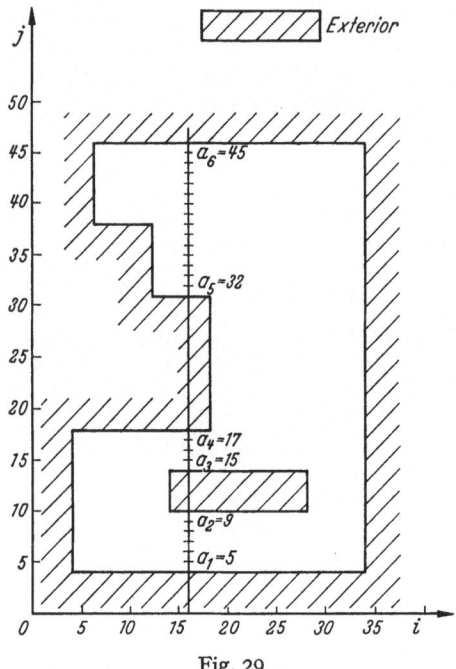

Fig. 29

Likewise a square domain subdivided into a grid of 10×10 smaller squares would appear as the degenerate **array** $a[1\!:\!1, -2\!:\!2]$:

1	9	2	1	9

To solve now the Dirichlet problem, e.g. by overrelaxation, we assume that the solution at grid point i, j is denoted by $x[i, j]$ and $f(x, y)$ at i, j by $f[i, j]$:

```
«begin
     for i := 0 step 1 until imax do
         for j := 0 step 1 until jmax do x[i, j] := 0 ;
     for k := 1 step 1 until kmax do
     begin
         rr := 0 ;
step:    for c := 1 step 1 until n do
             for i := a[c, −2] step 1 until a[c, −1] do
                 for p := 2 step 2 until a[c, 0] do
                     for j := a[c, p−1] step 1 until a[c, p] do
```

begin
$$res := 4 \times x[i, j] - x[i, j-1] - x[i, j+1]$$
$$- x[i+1, j] - x[i-1, j] - h\uparrow 2 \times f[i, j] \; ;$$
$$rr := rr + res\uparrow 2 \; ;$$
$$x[i, j] := x[i, j] - omega \times res/4$$
end c, i, p, j ;
    if $rr < eps\uparrow 2$ **then goto** *out*
    end k ;
out:
end ».

Statement *step*, which describes essentially one single overrelaxation step, is a fourfold loop: c counts the groups of vertical grid lines having identical intersections with L, i counts the vertical lines within the groups, p counts the sections which L cuts out of one vertical line in each group, and j counts the grid points within these sections.

§ 34. Least Square Problems

34.1. Orthogonalisation

The solution of more than m linear equations with m unknowns,

$$r_i = \sum_{k=1}^{m} a_{ik} x_k + b_i = 0 \qquad (i = 1, 2, \ldots, n > m)$$

in the sense of Gauss is characterized by

$$\sum_{i=1}^{n} r_i^2 = \min.$$

and is found most conveniently by orthogonalisation of the columns of the matrix $A = (a_{ik})$. The Schmidt orthogonalisation process requires strict orthogonality of the resulting vectors, but this is not guaranteed automatically if the process is carried out numerically. Indeed, if the orthogonalisation step

$$\vec{v}_k^{(old)} \to \vec{v}_k^{(new)} = \vec{v}_k^{(old)} - \sum_{j=1}^{k-1} r_{kj} \vec{v}_j^{(new)} \qquad (1)$$

(see also statement *orth* in the program below) produces a vector $\vec{v}_k^{(new)}$ which is much shorter than the given $\vec{v}$ was, then the roundoff errors may have the effect that $\vec{v}_k$ is oblique to some of the vectors $\vec{v}_1, \vec{v}_2, \ldots, \vec{v}_{k-1}$. To avoid this, the orthogonalisation is repeated (statement *rep*) whenever the reduction (1) reduces the length $\sqrt{tt}$ of $\vec{v}_k$ to less than one tenth[1] of

---

[1] The program will fail if the length of the vector $\vec{v}_k$ is reduced to zero.

its original length, but the $r[i, k]$ produced by the repetition have no further meaning:

```
«begin
     for k := 1 step 1 until m do
     begin
          t := tt := 0 ;
          for j := 1 step 1 until n do t := t + a[j, k]↑2 ;
orth:     for i := 1 step 1 until k − 1 do
          begin
               s := 0 ;
               for j := 1 step 1 until n do s := s + a[j, i] × a[j, k] ;
               if tt = 0 then r[i, k] := s ;
               for j := 1 step 1 until n do a[j, k] := a[j, k] − s × a[j, i] ;
          end i ;
          tt := 0 ;
lth:      for j := 1 step 1 until n do tt := tt + a[j, k]↑2 ;
          if tt < 0.01 × t then
rep:      begin
               t := tt ;
               goto orth
          end if ;
          r[k, k] := sqrt (tt) ;
norm: for j := 1 step 1 until n do a[j, k] := a[j, k]/r[k, k]
     end k
end».
```

The result of this algorithm is a new $n \times m$ matrix $A^{(\text{new})}$, which is stored in place of the given matrix A, and an upper triangular matrix $R = (r_{ik})$ such that $A^{(\text{new})} \times R$ equals the given matrix A.

To solve now the problem $|A \vec{x} + \vec{b}| = \text{Min.}$, we add $\vec{b}$ to A as the $m+1$-th column and apply the above program to the extended $n \times (m+1)$ matrix (to this effect we must use the program with $m+1$ in place of m). Then the components of $\vec{x}$ are the solution to the linear system

$$
\begin{aligned}
r_{11} x_1 + r_{12} x_2 + \cdots + r_{1,m} x_m + r_{1,m+1} &= 0 \\
r_{22} x_2 + \cdots + r_{2,m} x_m + r_{2,m+1} &= 0 \\
&\ \ \vdots \\
r_{m,m} x_m + r_{m,m+1} &= 0,
\end{aligned}
$$

whereas $r_{m+1,m+1}$ is the attained minimal value of $|A \vec{x} + \vec{b}|$.

34.2. Generation of orthogonal polynomials

Following G. FORSYTHE [14] we can generate polynomials $P_k(x)$ which are orthogonal with respect to a given weight-function $w(x)$, i.e. such that

$$\int_a^b P_i(x) P_j(x) w(x) \, dx = \delta_{ij},$$

by the tri-term recurrence relation

$$b_k P_k(x) = (x - a_k) P_{k-1}(x) - b_{k-1} P_{k-2}(x),$$

where

$$a_k = \int_a^b x w(x) P_{k-1}^2(x) \, dx,$$

and b_k is determined such that

$$\int_a^b w(x) P_k^2(x) \, dx = 1.$$

The program given below assumes that the orthogonality interval is $(0, 1)$ and that $w(x)$ is given in tabular form as a vector $w[0:n]$ where

$$w[j] = \begin{cases} \dfrac{1}{2n} w(j/n) & \text{for} \quad j = 0, n \\[2mm] \dfrac{1}{n} w(j/n) & \text{for} \quad j \neq 0, n. \end{cases}$$

It computes all integrals by the trapezoidal rule and produces also the polynomials $P_k(x)$ in tabular form:

```
«begin
    s := 0 ;
    for j := 0 step 1 until n do s := s + w[j] ;
    s := 1/sqrt(s) ;
    for j := 0 step 1 until n do p[0, j] := s ;
    for k := 1 step 1 until m do
    begin
comment Here P_{k-1}(x) and for k ≠ 1 also P_{k-2}(x) are available and
        normalized ;
    s := 0 ;
    for j := 0 step 1 until n do s := s + j × p[k − 1, j]↑2 × w[j] ;
    a[k] := s/n ;
comment Generate Polynomial P_k(x) ;
    if k = 1 then for j := 0 step 1 until n do
        p[k, j] := (j/n − a[k]) × p[k − 1, j] ;
    if k ≠ 1 then for j := 0 step 1 until n do
        p[k,j] := (j/n − a[k]) × p[k − 1, j] − b[k − 1] × p[k − 2, j] ;
    s := 0 ;
```

comment Normalize $P_k(x)$;
    for $j := 0$ **step** 1 **until** n **do** $s := s + w[j] \times p[k, j] \uparrow 2$;
    $b[k] := sqrt(s)$;
    for $j := 0$ **step** 1 **until** n **do** $p[k, j] := p[k, j]/b[k]$
  end k ;
end ».

At the end, the orthogonal polynomials $P_k(x)$ $(k = 0, 1, \ldots, m)$ are available in tabular form as the rows of the matrix p, $p[k, j]$ being approximately the value of $P_k(x)$ at the meshpoint $x = j/n$.

To use these polynomials for curve fitting, i.e. to approximate $f(x)$ by $\sum_0^m c_k P_k(x)$ such that

$$\int_0^1 w(x) \left(f(x) - \sum_0^m c_k P_k(x)\right)^2 dx = \text{Min.},$$

the c_k can be determined as follows (It is assumed that $f(x)$ is a function designator which computes the value of f at x):

«**for** $k := 0$ **step** 1 **until** m **do**
 begin
  $s := 0$;
  for $j := 0$ **step** 1 **until** n **do** $s := s + w[j] \times f(j/n) \times p[k, j]$;
  $c[k] := s$
 end ».

34.3. Chebychev series development

Let $f(x)$ be a function defined on the interval $(-1, 1)$. We require a least-square approximation

$$f(x) = \frac{c_0}{2} + \sum_{k=1}^{n-1} c_k T_k(x), \tag{2}$$

where $T_k(x) = \cos(k \times arccos(x))$, with $w(x) = 1/sqrt(1 - x\uparrow 2)$.

To obtain the c_k, we compute

where
$$c_k^{(n)} = \frac{2}{n} \sum_{j=0}^{n-1} f(x_j) \cos\left\{\frac{j+\frac{1}{2}}{n} k\pi\right\}, \left.\begin{array}{l} \\ \\ \end{array}\right\} k = 0, 1, \ldots, n-1. \tag{3}$$
$$x_k = \cos\left\{\frac{k+\frac{1}{2}}{n}\pi\right\}$$

These $c_k^{(n)}$ are computed for increasing values of n until they settle down to limits

$$c_k = \lim_{n \to \infty} c_k^{(n)},$$

which are the coefficients occurring in (2).

Since the values $cos\left\{\dfrac{j+\frac{1}{2}}{n}k\pi\right\}$ are used over and over again in the process, they are not computed using the standard function $cos(x)$; instead we generate for every new n a table consisting of the values $d_p = cos\left(\dfrac{p}{2n}\pi\right)$ $(p = 0, 1, 2, \ldots, 4n)$. Then we can read from this table the values $x_k = d_{2k+1}$ and $cos\left\{\dfrac{j+\frac{1}{2}}{n}k\pi\right\} = d_{k(2j+1)}$, where of course $k(2j+1)$ must be taken modulo $4n$.

If we let n run through the values 2, 4, 8, 16, ..., then we can compute the d's for $2n$ recursively from those for n by virtue of the half-argument formulae of the cosine. On the whole we obtain the program below. Termination occurs either by a jump to the label *noncon* (supposed to be somewhere outside this piece of program), if *nmax* was insufficient to yield sufficiently close agreement between the $c_k^{(n/2)}$ and $c_k^{(n)}$, or otherwise by an exit through «**end**». In the latter case the agreement may still be accidental, and in this respect this piece of program is not quite foolproof.

```
«begin
    for k := 0 step 1 until nmax do c[k] := 0 ;

comment Compute initial values of the d[k] (corresponding to n=1) ;
    d[0] := d[4] := 1 ;
    d[1] := d[3] := 0 ;
    d[2] := − 1 ;
    for n := 2, 2×n while n≤nmax do
    begin

comment Start computation of all c[k] for one n. Compute first new
        values d[k] interlaced between the old ones ;
        for j := 2×n step − 1 until 1 do d[2×j] := d[j] ;
        d1 := sqrt(2+2×d[2]) ;
        for j := 1 step 2 until 4×n− 1 do d[j] := (d[j+1]+d[j−1])/d1 ;
        t := 0 ;
        for j := 0 step 1 until n − 1 do fct[j] := f(d[2×j+1]) ;
        for k := 0 step 1 until n − 1 do
        begin

comment Start summation (3) for computing c[k] ;
            s := 0 ;
            jk := k ;
            for j := 0 step 1 until n − 1 do
            begin
```

comment jk is congruent $k \times (2 \times j + 1)$ modulo $4 \times n$;
$\qquad s := s + d[jk] \times fct[j]$;
$\qquad jk := jk + 2 \times k$;
$\qquad$ **if** $jk > 4 \times n$ **then** $jk := jk - 4 \times n$
$\qquad$ **end** j ;
$\qquad c1 := 2 \times s/n$;
$\qquad t := t + abs(c1 - c[k])$;
$\qquad c[k] := c1$;
$\qquad$ **end** k ;
comment t is sum of absolute differences between old and new $c[k]$'s ;
$\qquad$ **if** $t < eps$ **then** **goto** *limit* ;
$\qquad$ **end** n ;
$\qquad$ **goto** *noncon* ;
limit:
$\quad$ **end** ».

§ 35. Computations Related to Continued Fractions

35.1. Introduction

Continued fractions are constructions of the form

$$\frac{f_1 \mid}{\mid g_1} + \frac{f_2 \mid}{\mid g_2} + \frac{f_3 \mid}{\mid g_3} + \cdots + \frac{f_m \mid}{\mid g_m}, \tag{1}$$

where f_k and g_k are given functions of the subscript k. In analysis this concept is extended to $m = \infty$ and thus successfully used to represent various transcendental functions[1]. As an example, if $f_1 = 1$, $f_k = k - 1$ for $k > 1$, $g_k = z$ for all k, the (infinite) continued fraction converges for all positive real z to

$$e^{z^2/2} \int_z^{\infty} e^{-t^2/2} dt,$$

whereas the corresponding power series

$$\frac{1}{z} - \frac{1}{z^3} + \frac{3}{z^5} - \frac{3 \times 5}{z^7} + \frac{3 \times 5 \times 7}{z^9} - + \cdots$$

is an asymptotic expansion, diverging for all z. Likewise in many other instances divergent or badly convergent power series can be transformed into convergent continued fractions.

In numerical practice the continued fractions must of course always be truncated to finite length and then have the form (1). If the coefficients f, g are given as two arrays, (1) can be evaluated by *backward recurrence*, which bases on the fact that the relation $s^{(k)} = f_k/(g_k + s^{(k+1)})$ holds between the values

$$s^{(k)} = \frac{f_k \mid}{\mid g_k} + \frac{f_{k+1} \mid}{\mid g_{k+1}} + \cdots + \frac{f_m \mid}{\mid g_m} \quad (k = 1, 2, \ldots, m).$$

[1] See H. S. WALL [39].

Introducing $s^{(m+1)}=0$ and omitting the superscripts of s, we obtain the following piece of program which computes the value of (1) as cf:

«**begin**
    $s := 0$;
    for $k := m$ **step** -1 **until** 1 **do** $s := f[k]/(s+g[k])$;
    $cf := s$
end».

35.2. Evaluation by the forward recurrence relation

The above method of evaluation is usually not appropriate if we are concerned with truncated infinite continued fractions because then m is not known a priori but rather depends upon the speed of convergence. Moreover, it is in this case also uneconomical to assume the f_k, g_k as being given as components of two arrays. Therefore the following modifications are recommended: *First*, the f_k, g_k are computed for every k by function designators «$ff(k)$», «$gg(k)$». *Second*, we use the *forward recurrence* formulae[1]:

$$B_0=1, \quad A_0=0, \quad A_1=f_1, \quad B_1=g_1,$$

$$A_k=g_k A_{k-1}+f_k A_{k-2}, \quad B_k=g_k B_{k-1}+f_k B_{k-2}, \quad k=2,3,\ldots, \text{ad inf.} \quad (2)$$

with which the value of the (infinite) continued fraction is obtained as $\lim_{k\to\infty}(A_k/B_k)$ and can therefore be approximated to any desired accuracy.

However, there is some problem in storing the A_k, B_k: Since A_k depends on A_{k-1} and A_{k-2}, we cannot discard A_{k-1} after computation of A_k, but we can overwrite A_{k-2} with A_k (and likewise for the B's). This suggests the use of two positions for the A_k, say $a[1]$ for even, $a[-1]$ for odd k, and $b[1], b[-1]$ for B. Thus with the termination criterion $\left|\dfrac{A_k}{B_k}-\dfrac{A_{k-1}}{B_{k-1}}\right| < eps$ being checked every r-th step, the following program emerges (compare also 36.4):

«**begin**
    $p := q := 1$;
    $a[1] := 0$; $b[1] := 1$;
    $a[-1] := ff(1)$; $b[-1] := gg(1)$;
    for $k := 2$ **step** 1 **until** $kmax$ **do**
    begin
        $f := ff(k)$; $g := gg(k)$;
        $a[p] := g \times a[-p]+f \times a[p]$;
        $b[p] := g \times b[-p]+f \times b[p]$;
        if $k=q \times r$ **then**

---

[1] See H. S. WALL [39], § 1.

```
    begin
      cf := a[p]/b[p] ;
      cg := a[-p]/b[-p] ;
      if abs(cf - cg) < eps then goto ex ;
      q := q + 1 ;
    end if ;
      p := -p ;
  end k ;
ex:
  end».
```

35.3. Transformation of a power series into a continued fraction

To "nearly every" power series $\sum_{k=0}^{\infty} c_k x^k$ there exists a corresponding continued fraction

$$\frac{c_0}{|1} - \frac{a_1 x}{|1} - \frac{a_2 x}{|1} - \cdots$$

which is uniquely defined by the property that for every m the rational function

$$f_m(x) = \frac{c_0}{|1} - \frac{a_1 x}{|1} - \frac{a_2 x}{|1} - \cdots - \frac{a_m x}{|1} \tag{3}$$

agrees up to the x^m-term with the given power series.

One method for computing the coefficients a_k from given c_k is the quotient-difference algorithm [28], which can be described as follows: From the given coefficients c_k we compute

the quotients $\qquad q_1^{(k)} = c_{k+1}/c_k$,

the differences $\qquad e_1^{(k)} = q_1^{(k+1)} - q_1^{(k)}$,

the quotients $\qquad q_2^{(k)} = (e_1^{(k+1)}/e_1^{(k)}) \times q_1^{(k+1)}$,

the differences $\qquad e_2^{(k)} = (q_2^{(k+1)} - q_2^{(k)}) + e_1^{(k+1)}$ etc.,

generally $\qquad \begin{cases} e_j^{(k)} = q_j^{(k+1)} - q_j^{(k)} + e_{j-1}^{(k+1)} \\ q_{j+1}^{(k)} = (e_j^{(k+1)}/e_j^{(k)}) \times q_j^{(k+1)}. \end{cases}$

(putting $e_0^{(k+1)} = 0$) $\hspace{6em}$ (4)

Then the coefficients a_k of the corresponding continued fraction are

$$\begin{aligned} a_{2k-1} &= q_k^{(0)} \\ a_{2k} &= e_k^{(0)} \end{aligned} \right\} \quad (k = 1, 2, \ldots).$$

The rules (4) are transformed into an easily memorizable form (so-called rhombus rules, cf. E. STIEFEL [35]) if we arrange the q's and e's in a quotient-difference table:

$$
\begin{array}{cccccccc}
c_0 \\
& q_1^{(0)} \\
c_1 && e_1^{(0)} \\
& q_1^{(1)} && q_2^{(0)} \\
c_2 && e_1^{(1)} && e_2^{(0)} \\
& q_1^{(2)} && q_2^{(1)} && q_3^{(0)} \\
c_3 && e_1^{(2)} && e_2^{(1)} && \vdots \\
\vdots & q_1^{(3)} & \vdots & q_2^{(2)} & \vdots & \ddots \\
& \vdots && \vdots
\end{array}
$$

In this two-dimensional array the c's in the leftmost column are given and the $q_k^{(0)}$, $e_k^{(0)}$ in the top diagonal row are sought. They can be computed by systematic application of the rhombus rules from left to right. To obtain an economical program, we must find an arrangement which does not require all these elements to be stored at the same time, but if possible only one row or column at once. To do this, let us first rename the entries of the qd-table as follows:

$$
\begin{array}{cccccccc}
c_0 \\
& a_1^{(1)} \\
c_1 && a_2^{(2)} \\
& a_3^{(2)} && a_3^{(3)} \\
c_2 && a_4^{(3)} && a_4^{(4)} \\
& a_5^{(3)} && a_5^{(4)} && a_5^{(5)} \\
c_3 && a_6^{(4)} && a_6^{(5)} && \vdots \\
\vdots & a_7^{(4)} & \vdots & a_7^{(5)} & \vdots & \ddots \\
& \vdots && \vdots
\end{array}
$$

For every new c_k given, we compute one new antidiagonal consisting of the values $a_{2k-1}^{(k)}, a_{2k-2}^{(k)}, \ldots, a_k^{(k)}$ (in this order), where $a_p^{(k)}$ is computed from $a_p^{(k-1)}, a_{p-1}^{(k-1)}, a_{p+1}^{(k)}$ according to (4). We observe that $a_p^{(k-1)}$ is no longer used after $a_p^{(k)}$ has been computed; therefore the former can be overwritten by the latter, which means that the quantities $a_p^{(k)}$ can be stored for all k as the same array component $a[p]$. In the following piece of program the rhombus rules are distinguished by a Boolean variable q ($q =$ **true** refers to the Q-rule, $q =$ **false** to the E-rule).

```
«for k := 1 step 1 until m do
  begin
    a[2×k−1] := c[k]/c[k−1] ;
    a[2×k−2] := 0 ;
    q := true ;
    for j := 2×k−2 step −1 until k do
```

```
begin
    comment rhombus rules ;
    a[j] := if q then a[j+1]+a[j]-a[j-1]
             else a[j+1]×a[j]/a[j-1] ;
    q := ¬q
  end j ;
end k».
```

After execution of this program, the $a_k = a[k]$ $(k = 1, 2, \ldots, n)$ are the coefficients of the finite continued fraction (3). In other words, the $a_1 \ldots a_m$ are the first m coefficients of the infinite continued fraction which corresponds to the power series $\sum\limits_{k=0}^{\infty} c_k x^k$. The coefficients $a_{m+1}, \ldots, a_{2m-1}$ on the other hand are not further used.

35.4. The epsilon algorithm[1]

If we intend to compute the value of (3) for many values of x, then it is appropriate to compute first the coefficients $a[1:m]$ and then evaluate for every x the finite continued fraction (3). However, if we want to evaluate $f_m(x)$ just for one x, then it is computed more economically by the epsilon algorithm. This is a method which derives from a given sequence $eps_0^{(0)}, eps_0^{(1)}, eps_0^{(2)}, \ldots$, a two-dimensional array $eps_i^{(k)}$, usually arranged as

$$
\begin{array}{cccccccc}
eps_0^{(0)} & & & & & & \\
& eps_1^{(0)} & & & & & \\
eps_0^{(1)} & & eps_2^{(0)} & & & & \\
& eps_1^{(1)} & & eps_3^{(0)} & & & \\
eps_0^{(2)} & & eps_2^{(1)} & & \ddots & & \\
& eps_1^{(2)} & & eps_3^{(1)} & & & \\
eps_0^{(3)} & \vdots & eps_2^{(2)} & \vdots & & \\
\vdots & & \vdots & & & \\
\end{array}
$$

According to P. WYNN [41], the following is true: If $\lim\limits_{k\to\infty} eps_0^{(k)}$ exists, then usually also the sequence $eps_{2j}^{(0)}, eps_{2j}^{(1)}, eps_{2j}^{(2)}, \ldots$ converges for every j to the same limit, and so does the diagonal sequence $eps_0^{(0)}, eps_2^{(0)}, eps_4^{(0)}, eps_6^{(0)}, \ldots$ Usually the derived sequences converge faster than the given one, and sometimes they converge even if the given sequence diverges.

To adapt the method for practical requirements, we condense the epsilon array to those columns with even subscripts since only those

---

[1] P. WYNN [41].

approximate the limit of the given sequence:

$$
\begin{array}{llll}
eps_0^{(0)} \\
eps_0^{(1)} & eps_2^{(0)} \\
eps_0^{(2)} & eps_2^{(1)} & eps_4^{(0)} \\
eps_0^{(4)} & eps_2^{(2)} & eps_4^{(1)} & eps_6^{(0)} \\
\;\;\vdots & \;\;\vdots & \;\;\vdots & \;\;\vdots
\end{array}
$$

For the condensed epsilon-table we have also condensed recurrence relations (P. WYNN [42]):

$$
\frac{1}{C-W} + \frac{1}{C-E} = \frac{1}{C-S} + \frac{1}{C-N}, \tag{5}
$$

where C, N, W, S, E are any five elements of the condensed array standing in the relative positions

$$
\begin{array}{ccc}
 & N & \\
W & C & E \\
 & S &
\end{array}
$$

Relation (5) is also true if C is an element of the leftmost column, provided we set $W = \infty$ (we shall use $_{10}30$ instead). With this extension, (5) is sufficient for building up the condensed array from the given values $eps_0^{(k)}$ simply by solving the relation for E and letting C run through all entries except the top diagonal of the array.

However, also here we must avoid having to retain all elements of the array simultaneously in storage. To achieve this, we rename the condensed array as follows:

$$
\begin{array}{lll}
eps[0] \\
eps[1] & eps[0] \\
eps[2] & eps[1] & eps[0] \\
eps[3] & eps[2] & eps[1] \\
eps[4] & eps[3] \\
eps[5]
\end{array}
$$

and assume that at a given moment the last two elements of every column (the elements below the dotted line) are available. Obviously, as soon as the next element $eps[k]$ (here $k=6$) of the first column is given, the recursion rule (5) can be applied and a new element can be computed in all other columns; moreover, a new column is started if k is even. At the same time we can overwrite the elements immediately below the dotted line, which completes the step from $k-1$ to k.

However, one difficulty remains: the new elements are computed before their respective storage positions become available. Indeed, if e.g. the next element $eps[2]$ in the third column is computed, the old $eps[2]$ is still needed for computing a new $eps[0]$ in the fourth column, but $eps[4]$ can be overwritten. For this reason we store the new elements generally as $eps[j+2]$ instead of $eps[j]$ and shift them back at the end of every step. This we do also for the given $eps[k]$. All in all we obtain the following piece of program:

```
«begin
    for k := kmax step −1 until 0 do eps[k+2] := eps[k] ;
    jj := 1 ;
    for k := 0 step 1 until kmax do
    begin
        jj := 1−jj ;
comment At this point jj is congruent k modulo 2. jj serves to distin-
        guish cycles which produce epsilons with even subscripts
        from cycles in which epsilons with odd subscripts are com-
        puted ;
        eps[k] := 10 30 ;
        for j := k−2 step −2 until 0 do eps[j+2] := eps[j+1] −
            1/(1/(eps[j+1] − eps[j]) + 1/(eps[j+1] − eps[j+4]) −
            1/(eps[j+1] − eps[j+2])) ;
        for j := jj step 2 until k do eps[j] := eps[j+2]
    end k
end».
```

After termination $eps[jj]$ will (usually) be the best approximation for $\lim_{k \to \infty} eps_0^{(k)}$; in fact, for $kmax =$ even, and if the $eps_0^{(k)}$ are the partial sums of a power series $\sum_0^\infty c_j x^j$, then $eps[0]$ coincides (theoretically) with the value that would have resulted by evaluating the finite continued fraction (3) with $m = kmax − 1$.

§ 36. Considerations Concerning Computer Limitations

The designer of an ALGOL program for a typical textbook algorithm will observe that it takes comparatively little effort to transcribe the algorithm into ALGOL, but he will also observe that it often requires a much greater additional effort to achieve a program that produces useful results despite roundoff errors and other computer limitations. The present section serves to show some of the countermeasures which must be built into programs in order to make them run properly and — last but not least — to ensure proper termination of the program.

36.1. Quadratic equations

On first sight one would hardly suspect that the classic formula $x := -p/2 + sqrt(p\uparrow2/4 - q)$ for solving the equation $x^2 + px + q = 0$ might be endangered by computer limitations, and yet it bears two sources of trouble:

First, if we solve e.g. $x^2 - 700x + 1 = 0$ with 7 decimal digits relative precision, the above formula yields for the smaller root:

$$x = 350 - sqrt(122499) = 350 - 349.9986 = 0.0014.$$

Due to cancellation of digits, this result has a relative error of 2 %. This rather poor result can of course be improved by computing the larger root first, after which $x_2 = q/x_1$; this yields here

$$x_2 = 1/699.9986 = 0.001428574.$$

Second, the above formula, because of the occurrence of $p\uparrow2$, can be used only for about half the exponent range. Indeed, in a computer in which floating point numbers are confined to the interval

$$x < 2^{64} = 1.844 \cdot 10^{19}, \quad \text{the equation} \quad x^2 - 10^{12}x + 10^{16} = 0$$

can no longer be solved by that formula and this despite the fact that the roots (10^{12} and 10^4 approximately) are well within the prescribed range.

The following program for computing the two roots can be used in the full number range and without fear of inexactness for the smaller root *x2*:

```
«begin
    if abs (p) > 100 then
    begin
        d := 1/4 - q/p/p ;
        if d < 0 then goto complx ;
        x1 := - p × (1/2 + sqrt (d)) ;
    end
    else
        begin
            d := p↑2/4 - q ;
            if d < 0 then goto complx ;
            x1 := - p/2 - (if p > 0 then sqrt (d) else - sqrt (d))
        end if - else ;
    x2 := if x1 = 0 then 0 else q/x1
end».
```

Note. The somewhat strange expression $1/4 - q/p/p$ on the fourth line of the program would seem to be equivalent to $1/4 - q/p\uparrow2$; however, the latter implies evaluation of $p\uparrow2$, which might again cause overflow, whereas $q/p/p$ would not (it might produce underflow, but this is no problem).

36.2. Newton's method

In order to improve the unsatisfactory jumpout condition of the program given in 31.2, it should be recognized that the true source of the trouble are the roundoff errors involved in the computation of $f(x)$, which completely overshadow the value of f as soon as it becomes small. A possible remedy is therefore to jump out of the loop as soon as f comes down to the order of magnitude of the roundoff errors. This requires that the influence of these errors (in the computation of f) be carefully estimated:

The essence of computing f is the recurrence relation $f_{new} := f_{old} \times x + a[k]$ in which we distinguish for the moment the value of f before and after the operation. By this formula the error of f_{old} is multiplied by x, but also a new error is produced, namely, if h denotes the largest possible relative error of the computer:

$\leqq h \times x \times f_{old}$ in the multiplication $x \times f_{old}$,

$\leqq h \times Max(|x \times f_{old}|, |a[k]|)$ in the adjustment (if any) before the addition of $a[k]$,

$\leqq h \times f_{new}$ in the adjustment (if any) after addition of $a[k]$.

Observing that $Max(|x \times f_{old}|, |a[k]|) \leqq |x \times f_{old}| + |f_{new}|$, the error contributions can be computed parallel to the computation of f itself in the same loop as follows (a variable *noise* is introduced which is to be multiplied by h to obtain the maximum error itself):

$$
\begin{aligned}
&\text{«} noise := f := 0 \,; \\
&\left.
\begin{aligned}
noise &:= abs(x) \times noise + abs(x \times f) \,; \\
f &:= x \times f + a[n] \,; \\
noise &:= noise + abs(f) \,;
\end{aligned}
\right\} k = n \\[4pt]
&\left.
\begin{aligned}
noise &:= abs(x) \times noise + abs(x \times f) \,; \\
f &:= x \times f + a[n-1] \,; \\
noise &:= noise + abs(f) \,;
\end{aligned}
\right\} k = n-1 \\[4pt]
&\left.
\begin{aligned}
noise &:= abs(x) \times noise + abs(x \times f) \,;
\end{aligned}
\right\} k = n-2 \\
&\qquad\qquad\qquad \vdots \\[4pt]
&\left.
\begin{aligned}
f &:= x \times f + a[0] \,; \\
noise &:= noise + abs(f) \,; \text{»}
\end{aligned}
\right\} k = 0
\end{aligned}
$$

This scheme shows that we can combine the operation $noise := noise + abs(f)$ at the end of one loop with the operation $noise := abs(x) \times noise + abs(x \times f)$ at the beginning of the next turn of the loop and replace it by $noise := abs(x) \times noise + 2 \times abs(x \times f)$. The final operation $noise := noise + abs(f)$, which is in this way omitted, is an unimportant contribution.

By other considerations we find that the iteration should be continued as long as the computed f remains above six times the noise level, i.e. as long as

$$abs(f) > 6 \times noise \times h.$$

However, this criterion uses a value h which is different for different computers. In order to arrive at a computer-independent criterion, we rewrite the above condition as

$$abs(f)/(6 \times noise) > h,$$

which is equivalent to

$$1 + abs(f)/(6 \times noise) > 1.$$

On the whole the following program is obtained:

```
«begin
r:    f := g := noise := 0 ;
      for k := n step −1 until 0 do
      begin
          noise := abs(x) × noise + 2 × abs(x × f) ;
          g := x × g + f ;
          f := x × f + a[k]
      end k ;
      x := x − f/g ;
      if 1 + abs(f)/(6 × noise) ≠ 1 then goto r
end».
```

Of course this program is not quite foolproof either, since we have still not eliminated the danger of a small or vanishing g, which should be banned, too. Also the imminent danger of overflow during calculation of f and g is not eliminated in the above program. Indeed, if we attempt to solve

$$x^{20} + 1000 x^{19} + 1 = 0,$$

starting with $x = -1000$, we obtain $f = 1$, $g = -10^{57}$, where the latter is already outside the (floating point) number range of certain computers. However, we do not pursue this problem any further here.

36.3. Monotonicity as a termination criterion

Where a theoretically monotonic iteration process is carried out numerically, it will be observed that monotonicity is lost after a certain number of steps. This is very likely, though not always, the proper moment to discontinue the process, which is all the more welcomed, as it is often difficult if not impossible to find other effective jumpout criteria for iteration processes.

A trivial but characteristic example is the iteration process

$$x_0 := (1+a)/2, \qquad x_{k+1} := (x_k + a/x_k)/2,$$

which produces a monotonically decreasing sequence converging to $\sqrt{a}$.

In actual computation, however, monotonicity is destroyed as soon as $\sqrt{a} - x_k$ comes down to the roundoff-error level; obviously this is the proper time for terminating the process.

To use the process in ALGOL (which is not actually needed because *sqrt* is available as standard function), we denote x_k by y, x_{k+1} by x and terminate as soon as $y > x$ (which theoretically should be **true** forever) no longer holds:

«**for** $x := (1+a)/2, \; (a/y+y)/2$ **while** $y > x$ **do** $y := x$».

Of course one might argue that $y > x$ could be true forever despite roundoff errors, in which case we would indeed obtain a closed loop. However, this cannot be so for the following reasons:

Assume that y agrees to more than half of the digits with a and that we have a binary computer, h having the same meaning as in 36.2. Then for certain θ's in the range $|\theta| \leq 1$ and omitting terms smaller than $O(h^{\frac{3}{2}})$:

$$y = \sqrt{a} + \theta_0 \sqrt{a} \sqrt{h}, \quad \text{hence} \quad a/y = \sqrt{a} - \theta_0 \sqrt{a} \sqrt{h} + \theta_0^2 \sqrt{a} h + \theta_1 h \sqrt{a}$$

(the last term being the roundoff error of division),

$$a/y + y = 2\sqrt{a} + \theta_0^2 \sqrt{a} h + \theta_1 h \sqrt{a} + 2\theta_2 \sqrt{a} h$$

(the last term being the roundoff error of the addition), and finally

$$x = \sqrt{a} + \theta_0^2 \sqrt{a} h/2 + \theta_1 h \sqrt{a}/2 + \theta_2 \sqrt{a} h = \sqrt{a} + 2\theta_3 \sqrt{a} h.$$

Now if $\sqrt{a}$ has a mantissa slightly below 2, then $2\sqrt{a}\, h$ are two units of the last place of x; the possible values for x must therefore lie on an interval whose length is four units of the last place, which leaves five possible values for x, and these same values are also possible for the following steps. Therefore, after at most four further steps, $y > x$ can no longer be true.

36.4. Overflow in continued fraction evaluation

If convergence of a continued fraction

$$\frac{f_1|}{|g_1} + \frac{f_2|}{|g_2} + \frac{f_3|}{|g_3} + \cdots$$

is slow, then the evaluation algorithm described in 35.2 must run up to very high values of k in order to achieve the desired result. However, in doing so, very large values A_k and B_k may be produced, causing overflow of the exponent of floating point number representation, even if the coefficients f_k, g_k are not large[1].

In order to prevent overflow as well as underflow it is recommended to check from time to time the size of A_k, B_k, A_{k-1}, B_{k-1} (which in fact are the values $a[1]$, $b[1]$, $a[-1]$, $b[-1]$ in our program) and rescale them whenever needed. This is done by the following program which checks every tenth step and to this end uses a subordinate (j-) loop for counting the ten steps inside the (k-) loop for counting the blocks of ten steps.

```
«begin
    p := 1 ;
    a[1] := 0 ;   b[1] := 1 ;
    a[-1] := ff(1) ;   b[-1] := gg(1) ;
    for k := 0 step 10 until kmax do
    begin
        for j := 2 step 1 until 11 do
        begin
            f := ff(j+k) ;   g := gg(j+k) ;
            a[p] := g×a[-p]+f×a[p] ;
            b[p] := g×b[-p]+f×b[p] ;
            p := -p
        end j ;
        cf := a[1]/b[1] ;
        cg := a[-1]/b[-1] ;
        if abs(c-cg) < eps then goto ex ;
        max := fabs(a[1])+abs(a[-1])+abs(b[1])+abs(b[-1]) ;
        if max > 10 20 then d := 10 -20
        else
            if max < 10 -20 then d := 10 20
            else
                goto out ;
        a[1] := d×a[1] ;   b[1] := d×b[1] ;
        a[-1] := d×a[-1] ;   b[-1] := d×b[-1] ;
out: end k ;
ex:
    end».
```

---

[1] The reader will observe that the bisection method for computing eigenvalues of tridiagonal matrices uses similar recurrence formulae and therefore is subject to the same danger of overflow.

36.5. Underflow in orthonormalisation processes

Schmidt orthonormalisation as described in 34.1 by formula (1) requires that the vectors $\vec{v}_1^{(new)}$, $\vec{v}_2^{(new)}$, ..., $\vec{v}_{k-1}^{(new)}$ be strictly orthonormal. In an attempt to guarantee this, special measures have been built into the program given in 34.1, and yet the program may still produce wrong results because of underflow that might occur during execution of the statement *lth*.

On first sight such underflow would seem extremely unlikely or even impossible; however, it should not be overlooked that if a large number of nearly parallel vectors should be orthonormalized, then every execution of statement *orth* may reduce the length of the k-th column vector of the array a considerably. Therefore, if the computer has a comparatively small exponent range, it may well occur that at a certain stage in the process some of the $a[j, k]\!\uparrow\!2$ (k fixed) are a little below and some just a bit above the underflow limit, whereupon the computed length of the k-th column vector of a is too small; hence the normalized k-th column is longer than 1 by an amount which is considerably larger than the roundoff errors would explain. Since the further course of the calculation requires (within computer accuracy) strict orthonormality, we must be prepared for erroneous results.

In order to avoid trouble of this sort, the following process is recommended for normalizing a vector $\vec{v}$ given by its components $v[1]$, $v[2], ..., v[n]$: We do not add the squares of $v[j]$ but instead the squares of $v[j]/vmax$, where $vmax$ denotes the maximum of all $abs(v[j])$. In this way the following program emerges:

```
«begin
    vmax := 0 ;
    for j := 1 step 1 until n do
        if abs(v[j]) > vmax then vmax := abs(v[j]) ;
    if vmax = 0 then goto zero ;
comment zero is the place where the computation continues in case v
        is the zero vector ;
    s := 0 ;
    for j := 1 step 1 until n do s := s + (v[j]/vmax)↑2 ;
    c := vmax × sqrt(s) ;
comment Here s = length of vector v ;
end».
```

Naturally the question of economy must be raised here, since this way of computing the length of a vector is an obvious waste of time if this length is not of extreme order of magnitude. Of course we can at least avoid the n divisions $v[j]/vmax$ if $\vec{v}$ is a vector of "normal" size, but for taking advantage of this possibility the size of *vmax* must be tested as follows:

«**if** (**if** *vmax* > 100 **then** 0.01/*vmax*/*vmax* **else** $_{10}$−10 × *vmax* × *vmax*) ≠ 0
  then begin comment Follows normal evaluation of length of *v* ;
  ... **end else begin comment** Abnormal case ; ... **end**».

In this if-clause the left hand side of the relation has the value 0.01/*vmax*↑2 or $_{10}$−10 × *vmax*↑2. The first alternative serves to exclude overflow, the second to prevent underflow in the computation of those *v*[*j*]↑2 which are still important contributions to *s* (assuming an 10-digit mantissa).

36.6. Bandmatrices

Another kind of computer limitation is the finite storage capacity which forces us to economization if large matrices are used in a calculation. A classical example of this sort are *bandmatrices*, which often occur in eigenvalue problems and as coefficient matrices of linear systems.

A bandmatrix is defined as one which has nonzero elements only in the vicinity of the diagonal, i.e. one whose elements $a[i, k]$ $(i, k = 1, 2, \ldots, n)$ have the property

$$a[i, k] = 0 \quad \text{for} \quad |i - k| > m,$$

where *m* is a certain number called the *bandwidth* of the matrix. Of course, according to this definition every matrix is a bandmatrix if we take $m = n - 1$, but the essence of this notion is that bandmatrices with *m* appreciably smaller than *n* allow enormous savings in computing time as well as storage requirements, if only the computation is organized properly.

In order to save storage space, the band of nonzero elements of the matrix $A = (a[i, j])$ must be transformed into a rectangular array $B = (b[i, k])$ as follows:

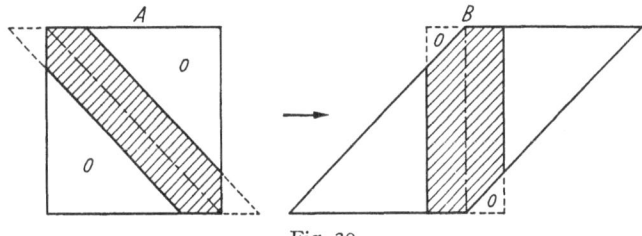

Fig. 30

The array element $a[i, j]$ of A is transformed into the element $b[i, j - i]$ of the array B, which, since $a[i, j] \neq 0$ only if $|j - i| \leq m$, can be declared as

«**array** $b[1:n, -m:m]$»

and thus requires only $(2m+1)n$ storage places instead of $n\uparrow 2$ for the usual matrix notation.

The *bandmatrix notation*, as we shall call this kind of representation of a bandmatrix, has the property that the first subscript indicates the line, whereas the second subscript is the distance from the diagonal. The diagonal elements thus appear as $b[i, 0]$, whereas $b[i, k]$ with positive (negative) k denotes a super (sub)-diagonal element. For symmetric bandmatrices we might achieve a further saving since in this case B need only be declared as

$$\text{«\textbf{array} } b[1:n, 0:m]\text{».}$$

Note that the array b contains two small corners consisting of elements $b[i, j]$ (those with $i+j>n$ and those with $i+j<1$) which do not correspond to elements within the matrix A; in the following we shall assume that these $b[i, j]$ are all zero.

Of course we can use bandmatrix notation only if we reformulate our numerical methods in terms of this notation. How this may be done is shown below for the Banachiewicz modification of the Gauss elimination process as described for full matrices in 31.1:

First, it must be investigated which of the step-until elements of the Banachiewicz program must be changed as a consequence of the bandform of the matrix: The i-loop is not touched, but the subscript k is restricted to the interval $i-m\leq k\leq i+m$ since outside we have $a[i, k]=0$, and also statement *sum* would then be void. Indeed, in this latter statement the subscript j is restricted to the interval $max(k-m, i-m)\leq j\leq min(k-1, i-1)$ since for other j's either $a[i, j]$ or $a[j, k]$ vanishes.

Second, any $a[p, q]$ will be replaced throughout by $b[p, q-p]$, and new subscripts $j=j_{old}-i$ and $k=k_{old}-i$ are introduced. Observing the subscript bounds for j_{old}, k_{old}, we obtain the following bounds for j, k (compare also Fig. 31):

$l1=max(1-i, -m)\leq k\leq m$	(second for-clause)
$l2=max(k-m, -m, 1-i)\leq j\leq l3=min(k-1, -1)$	(third for-clause)
$l1\leq j\leq -1$	(fourth for-clause).

With these modifications, the following program (without backsubstitution) emerges:

```
«for i := 2 step 1 until n do
  begin
      l1 := if 1−i > −m then 1−i else −m ;
      for k := l1 step 1 until m do
```

```
begin
    l2 := if k − m > l1 then k − m else l1 ;
    l3 := if k > 0 then − 1 else k − 1 ;
    t := b[i, k] ;
    for j := l2 step 1 until l3 do t := t + b[i, j] × b[j + i, k − j] ;
    b[i, k] := if k < 0 then − t/b[k + i, 0] else t ;
end k ;
    for j := l1 step 1 until − 1 do s[i] := s[i] + b[i, j] × s[i + j]
end i ; ».
```

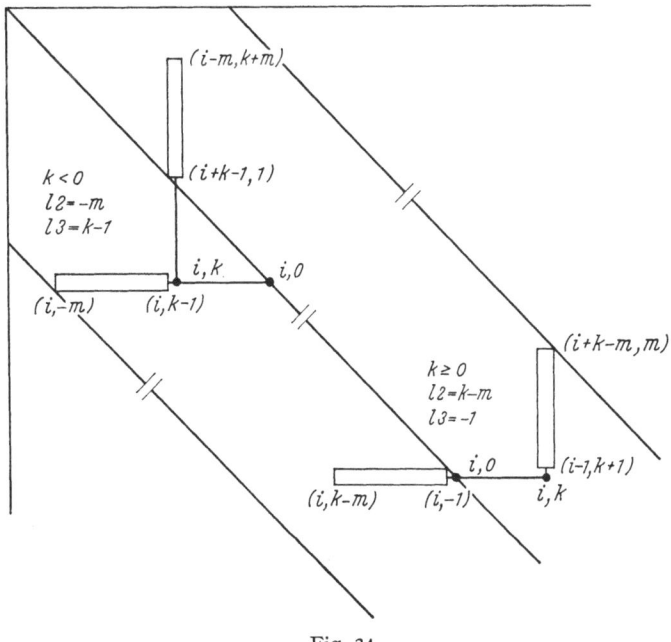

Fig. 31

According to the running of the subscripts, this piece of program solves a linear system with an effort on the order of nm^2 multiplications and additions compared to $n^3/3$ multiplications and additions for a full matrix. It is therefore well worth considering bandmatrix notation in such cases as $n = 200$, $m = 10$.

However, it should be recognized that this program, like the one given in 31.1, does not search for the largest pivot element and therefore can be used safely only under certain restrictions, e.g. if the coefficient matrix is symmetric and positive definite or diagonally dominant[1].

---

[1] Compare [38] for a program which solves linear systems in bandmatrix notation and yet searches for pivots.

§ 37. Data Processing Applications

Although ALGOL was not intended originally for general data processing, it seems that it can be used quite efficiently for this purpose. Of course, where data processing must be done with utmost efficiency, the use of ALGOL is not recommended. In fact, many processes described in this section can be executed far more efficiently by so-called code procedures (for which see § 47).

37.1. Pseudostring representation

It would appear that it is best to handle data in ALGOL as *strings* as they are described in § 12; however, since the operations that can be performed upon strings in ALGOL are entirely insufficient, we cannot use the string concept of ALGOL here. Instead we represent single characters by integers and therefore data, i.e. strings of characters, as integer arrays. As a consequence, all data processing operations must be simulated by operations upon integer arrays, and only upon input and output are these integer arrays converted from and into actual data (e.g. via the standard procedures *insymbol, outsymbol*, for which see § 50).

For the time being we make use of a fixed correspondence between ALGOL symbols and the integers 1 through 116:

Integer representation ↔ basic symbol

1	0	21	$=$	40	;	59	**array**	78	n	97	G
2	1	22	$\geqq$	41	:=	60	**switch**	79	o	98	H
3	2	23	$>$	42	⊔	61	**procedure**	80	p	99	I
4	3	24	$\neq$	43	**step**	62	**string**	81	q	100	J
5	4	25	**true**	44	**until**	63	**label**	82	r	101	K
6	5	26	**false**	45	**while**	64	**value**	83	s	102	L
7	6	27	$\equiv$	46	**comment**	65	a	84	t	103	M
8	7	28	$\supset$	47	(	66	b	85	u	104	N
9	8	29	$\vee$	48	)	67	c	86	v	105	O
10	9	30	$\wedge$	49	[	68	d	87	w	106	P
11	10	31	$\neg$	50	]	69	e	88	x	107	Q
12	.	32	**goto**	51	'	70	f	89	y	108	R
13	$+$	33	**if**	52	'	71	g	90	z	109	S
14	$-$	34	**then**	53	**begin**	72	h	91	A	110	T
15	$\times$	35	**else**	54	**end**	73	i	92	B	111	U
16	/	36	**for**	55	**own**	74	j	93	C	112	V
17	$\div$	37	**do**	56	**Boolean**	75	k	94	D	113	W
18	$\uparrow$	38	,	57	**integer**	76	l	95	E	114	X
19	$<$	39	:	58	**real**	77	m	96	F	115	Y
20	$\leqq$									116	Z

It should be recognized that this correspondence is in no way standardized or obliging but is only an ad hoc construction for our present purposes.

In the following we assume throughout that a string of basic symbols is represented as a *pseudostring*, i.e. an integer array, every component of which represents — by virtue of the correspondence table above — one single character. More precisely, a pseudostring has the form **integer array** $a[0:bigm]$, where $a[1], a[2], \ldots, a[a[0]]$ represent the characters which form the string, whereas $a[a[0]+1], \ldots, a[bigm]$ are meaningless. $a[0]$ is therefore the *actual* length of the string; $bigm$, however, is a standard upper bound for all strings to be used in the problem.

As an example, with $bigm = 80$,

$$state[0:80] = (9, 90, 41, 47, 86, 13, 2, 48, 18, 3, 0, \ldots, 0)$$

is the pseudostring representation of the ALGOL text «$z := (v+1)\uparrow 2$».

We can now describe the following elementary operations upon pseudostrings:

a) Append a pseudostring b to a pseudostring a. The length of the combined string is $a[0]+b[0]$:

«**begin**
    if $a[0]+b[0] > bigm$ **then goto** *overfl* ;
    for $k := 1$ **step** 1 **until** $b[0]$ **do** $a[k+a[0]] := b[k]$;
    $a[0] := a[0]+b[0]$
end».

b) Split a pseudostring a into two pseudostrings b, c of length p and $a[0]-p$ respectively. If p exceeds length $a[0]$ of given string a, c will be empty, hence $c[0]=0$, $b[0]=a[0]$.

«**begin**
    for $k := p+1$ **step** 1 **until** $a[0]$ **do** $c[k-p] := a[k]$;
    $c[0] :=$ **if** $p < a[0]$ **then** $a[0]-p$ **else** 0 ;
    $b[0] :=$ **if** $p < a[0]$ **then** p **else** $a[0]$;
    for $k := 1$ **step** 1 **until** $b[0]$ **do** $b[k] := a[k]$
end».

c) Compare two pseudostrings, disregarding spaces:

«**begin**
    $ka := kb := 0$;
xa: $ka := ka+1$;
    if $ka > a[0]$ **then goto** xb ;   String a exhausted.
    if $a[ka] = 42$ **then goto** xa ;   Skip space in a.
xb: $kb := kb+1$;
    if $kb > b[0]$ **then goto** xc ;   String b exhausted.
    if $b[kb] = 42$ **then goto** xb ;   Skip space in b.

xc: **if** $ka \leq a[0] \equiv kb > b[0]$ **then** *match* := **false**   The two strings are of different length.

    else

       if $ka > a[0] \wedge kb > b[0]$ **then** *match* := **true**   The two strings are identical.

       else

          if $a[ka] \neq b[kb]$ **then** *match* := **false**   A character pair in which the strings disagree is found.

          else

            goto xa ;   We have agreement up to this point; proceed to comparison of next character pair.

comment Here *match* = **true** iff the two strings agree ;

end».

37.2. Format handling

A value x of real type shall be printed with a simple *format* given as a sequence of 9's, sign, spaces, decimal point and arbitrary text (similar to the FORTRAN-F-format). As an example the format string «ⳑⳑxⳑ = ⳑ − 999.999 ⳑ 999 ⳑ ⳑ ⳑ» means that numbers should be printed e.g. as

$$x = \quad 003.141\ 592 \qquad \text{or} \qquad x = -752.002\ 357 \quad .$$

We do not attempt to suppress zeros or to round correctly at the end, but overflow will be signaled.

It is assumed that the format is given as a pseudostring f according to the above conventions and that the number to be printed will also be produced as such a pseudostring g (how the printing is actually done will be shown later in § 50). This task is performed by the following piece of program:

```
«begin
    for k := 1 step 1 until f[0] do
    begin
        if f[k] = 10 then x := x/10 ;
        if f[k] = 12 then goto out ;
    end k ;
out: g[0] := f[0] ;
    sign := (x ≥ 0) ;
    x := abs (x) ;
    if x ≥ 1 then goto overfl ;

    for k := 1 step 1 until f[0] do
    begin
        if f[k] = 14 then g[k] := if sign then 42 else 14
        else
            if f[k] = 13 then g[k] := if sign then 13 else 14
```

First pass (up to the decimal point of the format string) determines number of digits in the integral part and adjusts x.

{Given x exceeds size of format. Jump out of this program.

{Beginning of loop for generating actual output string g.

} Decision concerning printing of sign.

```
        else
          if f[k] = 10 then
          begin
              x := 10 × x ;
              g[k] := entier (x) + 1 ;
              x := x − entier (x)
          end
          else
              g[k] := f[k]
    end k
end».
```

$x := 10 \times x$; $g[k] := entier(x) + 1$; $x := x - entier(x)$	A 9 in the format causes printing of one digit and removal of the printed digit from the number x.
$g[k] := f[k]$	All other symbols are just copied.

37.3. Sorting

Let a be a two-dimensional integer array $a[1:n, 0:bigm]$, every row of which is a pseudostring of at most $bigm$ characters. These n sets of data should be sorted with respect to characteristics found in the s-th through t-th character of every pseudostring. In other words, the characters $a[k, s], a[k, s+1], \ldots, a[k, t]$ form the keyword for the k-th pseudostring; after sorting, all these keywords should appear in a prescribed order. For this purpose a function $\varphi(n)$ — in ALGOL given as a function designator $phi(n)$ — defines the ordering of the basic symbols such that the character represented by the integer p is considered as *preceding* the character represented by q if and only if $phi(p) < phi(q)$.

As to the technique of sorting, we assume that the first $k-1$ data sets (rows 1 through $k-1$ of the array a) have already been sorted. Then the correct place for the k-th data set among the $k-1$ already ordered sets is determined by binary search: Let p be the smallest power of 2 with the property $p \geq k$. Compare set k to set $p/2$, then depending upon whether set k comes before or after set $p/2$, compare set k to set $p/4$ or $3p/4$, etc. Of course $3p/4$ may exceed k; then we proceed in the same way as if set k were *before* (the nonexistent) set $3p/4$. As soon as we have compared set k to a set j with an odd value of j, the new position of set k is found.

Needless to say, we do not actually move the data in this sorting process but only record the actions which should be taken as a *permutation vector* $v[1:n]$, where $v[k]$ says that the $v[k]$-th row of array a should be in the k-th position.

```
«begin
    p := 2 ;
    v[1] := 1 ;
    for k := 2 step 1 until n do
```

```
      begin
        if k>p then p:= 2×p ;
        q:= l:= p/2 ;
xx:     first:= false ;
```

   comment Next statement makes $first =$ **true**, if and only if set k
            comes before set $v[l]$;

```
ww:     for j:= s step 1 until t do
          if phi(a[k,j]) ≠ phi(a[v[l],j]) then
          begin
            first:= phi(a[k,j]) < phi(a[v[l],j]) ;
            goto yy
          end if and j ;
yy:     if q ≠ 1 then
        begin
          q:= q/2 ;
          l := if first then l−q else l+q ;
          first:= true ;
          if l≥k then goto yy ;
          goto xx ;
        end if q ;
zz:     if ¬ first then l:= l+1 ;
vv:     if l≠k then
          for i:= k−1 step −1 until l do v[i+1]:= v[i] ;
        v[l]:= k
      end k
    end ».
```

Let us demonstrate the effect of this piece of program under the assumption that $s=t=1$ and that we have arrived at $k=12$, $p=16$, the 11 previous sets being ordered with $phi(a[v[i],1])=2i$, $(i= 1, 2, ..., 11)$, while $phi(a[12,1])=19$. We pass the following points in the program:

Beginning of k-loop: $k=12$, $p=16$, then $q:=8$; $l:=8$;

label xx: $first:=$ **false** ;
   ww: compare $phi(a[12,1])=19$ to $phi(a[v[8],1])=16$,
       $first:=$ **false** ;
   yy: $q:=4$; $l:=12$; $l≥k$, therefore **goto** yy !
   yy: $q:=2$; $l:=10$; $l<k$, **goto** xx !
   xx: $first:=$ **false** ;
   ww: compare $phi(a[12,1])=19$ to $phi(a[v[10],1])=20$,
       $first:=$ **true** ;
   yy: $q:=1$; $l:=9$; $l<k$, **goto** xx !

xx: $first :=$ **false** ;
ww: compare $phi(a[12, 1]) = 19$ to $phi(a[v[9], 1]) = 18$,
   $first :=$ **false** ;
yy: $q = 1$, hence proceed to
zz: $l := 10$;
vv: rearrange the v's: $v[10]$, $v[11]$ are shifted one place upwards,
   whereupon 12 is assigned to $v[10]$.

37.4. Differentiation of an arithmetic expression

Given a very simple arithmetic expression using only the arithmetic operators $+ - \times /$ and containing neither subscripted variables nor function designators, our task is to construct automatically another arithmetic expression which computes the derivative of the given expression with respect to a prescribed variable. This problem is of considerable practical importance since it can be extended to the transformation of an ALGOL program into another one which computes the derivatives of all results with respect to prescribed variables. However, only a rather restricted version of the general problem is solved here; as an example, we do not collect identical terms of the derived expression.

We assume that the arithmetic expression is given in pseudostring representation as «**integer array** $a[0:bigm]$» and also that the name of the variable with respect to which it should be differentiated is given as a pseudostring «**integer array** $name[0:bigm]$». Likewise the derived expression should be produced as a pseudostring $b[0:bigm]$.

In order to solve the restricted problem as stated before, we investigate the terms of the given expression E between adjacent additive operators $+$, $-$. These terms have the syntactic form

$$\text{«}V \ op \ V \ op \ V \ldots op \ V \ op \ V\text{»},$$

where the V's are numerical constants or variable identifiers and the op's denote the multiplicative operators $\times$, $/$. Every term is scanned, and for every occurrence of the variable X (with respect to which E should be differentiated) a new term of the derivative expression E' is produced in the obvious way. However, where X is preceded by a solidus $/$, the sign of the new term must be reversed and the sequence $/X\uparrow2$ must be inserted. Finally, the contributions of all terms are collected to form the expression E'.

In this process the following variables are used: s counts the characters read from E, t counts those put into E', $iden$ counts characters while an identifier is analyzed. During examination of a term of E, $op[0]$ indicates the position of the preceding additive operator within the pseudostring a, likewise $op[k]$ the position of the k-th multiplicative operator of the term. Finally, the Boolean variable $x[k]$ indicates whether or not the k-th factor of the current term is the variable X with respect to which we differentiate.

«begin

$$space := sig := t := iden := 0 ;$$

$\begin{cases} sig \text{ is sign of current term in} \\ \text{pseudostring notation. } sig = 0 \\ \text{means that only spaces have been} \\ \text{met so far.} \end{cases}$

for $s := 1$ **step** 1 **until** $a[0]$ **do**

Every turn of this loop analyses one character of E.

$l1:$ **begin**

   if $sig = 0 \wedge (a[s] = 13 \vee a[s] = 14)$ **then goto** $l11$
   else

$l2:$ **if** $sig = 0 \wedge a[s] \neq 42$ **then**
        begin
            $sig := 13$;
            $op[0] := s - 1$;
            $k := 0$;
            goto $l12$
        end if else ;

Initialisation; if E does not begin with a sign, a $+$ is placed in front of it.

   if $a[s] \geqq 13 \wedge a[s] \leqq 16$ **then**

$l3:$ **begin**
        $k := k + 1$;
        $op[k] := s$;
        if $iden < name[0]$ **then**
                    $match := $ **false** ;
        $x[k] := match$;
        $iden := 0$
   end $l3$;

If an arithmetic operator is found, a factor of a term is completed. k counts the factors in a term, $op[k]$ is position of k-th multiplying operator in a term, $op[0]$ denotes position of last adding operator. $x[k] = $ **true** means that the k-th factor of the term is the differentiation variable.

   if $a[s] = 13 \vee a[s] = 14$ **then**

$\begin{cases} \text{If an adding operator is found, a} \\ \text{term is completed; the derivative} \\ \text{of the term is now built up.} \end{cases}$

$l4:$ **begin**
       if $k = 1 \wedge x[1]$ **then**

$l5:$ **begin**
           $t := t + 2$;
           $b[t-1] := sig$;
           $b[t] := 2$
       end
       else

If there is only one factor and this is just the differentiation variable, the derivative is 1.

$l6:$ **for** $j := 1$ **step** 1 **until** k
               do if $x[j]$ **then**

Every occurrence of the differentiation variable X in a term gives rise to a term of E'. All these derived terms are now collected.

$l7$: **begin**

$t := t + 1$;

$b[t] := \textbf{if } a[op[j-1]] \neq 16$

then sig **else** $27 - sig$;

Occurrence of the differentiation variable as denominator causes a sign inversion.

$l8$: **for** $i := op[0] + 1$ **step** 1

until $op[j-1] - 1$ **do**

begin $t := t + 1$;

$b[t] := a[i]$

end ;

Copy that part not involved in differentiation.

if $a[op[j-1]] = 16$ **then**

$l9$: **begin**

for $i := op[j-1]$ **step** 1 **until**

$op[j] - 1$ **do**

if $a[i] \neq 42$ **then**

begin $t := t + 1$;

$b[t] := a[i]$

end ;

$t := t + 2$;

$b[t-1] := 18$;

$b[t] := 3$;

end $l9$;

If X occurs in a position $/X$, then $/X{\uparrow}2$ must be inserted; spaces are rearranged.

if $j = 1 \wedge a[op[j]] = 16$ **then**

begin

$t := t + 2$;

$b[t-1] := 2$;

$b[t] := 16$

end if j ;

Special handling if first factor is differentiation variable followed by $/$.

$l10$: **for** $i := (\textbf{if } j = 1 \textbf{ then } op[j] + 1$

else $op[j])$

step 1 **until** $op[k] - 1$ **do**

begin $t := t + 1$;

$b[t] := a[i]$ **end** ;

Copying where no differentiation is involved.

end $l7$;

if $x[k]$ **then**

for $i := 1$ **step** 1 **until** $space$ **do**

begin $t := t + 1$;

$b[t] := 42$ **end** ;

Special measures which copy spaces following a term into the derived expression also in that case where the term was differentiated with respect to its last factor.

l11:   **if** $s = a[0]$ **then goto** *ex* ;   Expression E exhausted.

      $sig := a[s]$;   ⎫ Initialisation of next term

      $k := 0$;   ⎪ (if any). $op[0]$ indicates po-

      $op[0] := s$;   ⎬ sition of adding operator in

      end *l4* ;   ⎭ front of a term.

l12:   **if** $a[s] \geq 65 \wedge a[s] \leq 90 \vee a[s] \leq 10 \wedge iden \neq 0$ **then**

      begin

         if $iden = 0$ **then** $match :=$ **true** ; ⎫ If a letter or a nonleading

         $iden := iden + 1$;   ⎪ digit is read, this is part of

         $match := match \wedge$ ⎬ an identifier which is now

                $(a[s] = name[iden])$;   ⎪ compared with the differ-

      end ;   ⎭ entiation variable X.

      if $s = a[0]$ **then**

l13:   **begin**

         $k := k + 1$;   ⎫ If last character of E has

         if $iden < name[0]$ **then**   ⎪ been read, the same mea-

                $match :=$ **false** ; ⎬ sures must be taken as if an

         $x[k] := match$;   ⎪ adding operator had been

         $op[k] := s + 1$;   ⎪ found.

         goto *l4*   ⎭

      end *l13* ;

      $space :=$ **if** $a[s] = 42$ **then** $space + 1$ **else** 0 ;

ex: **end** s ;

                                  If differentiated expression

      if $t = 0$ **then** $t := b[1] := 1$;   ⎱ is empty, we set $E' = 0$.

      $b[0] := t$ ⎰ Length of string E' is in-

                                  serted in $b[0]$.

  end ».

If the term $a \times b/c \times x/f$ (with $name = x$) is handled with this program, the statement *l2* first puts (for $s = 1$) $op[0] := 0$, $sig := 13$, and later yields:

$$
\left.
\begin{array}{ll}
x[1] = \textbf{false} & op[1] = 2 \\
x[2] = \textbf{false} & op[2] = 4 \\
x[3] = \textbf{false} & op[3] = 6 \\
x[4] = \textbf{true} & op[4] = 8
\end{array}
\right\}
\begin{array}{c}
\text{with statements } l3, l12 \\
\text{for } s = 1, 2, \ldots, 8.
\end{array}
$$

$x[5] = \textbf{false} \quad op[5] = 10$, with statement *l13*, for $s = 9$.

After that, $s = 10$, and the jump to *l4* is activated, which causes execution of statement *l7* with $j = 4$. Hereby the pseudostring b is built up: $b[1]$ is set to 13 (+), after which statement *l8* adds $a \times b/c$, and finally statement *l10* appends $/f$ to this pseudostring. Accordingly the result is

$$+ a \times b/c/f$$

in pseudostring notation.

37.5. Operations performed upon packed data

We have thus far assumed that every component of a pseudostring occupies a full machine word, but since these components are all positive numbers not exceeding 127, this is an obvious waste of storage space. To save storage, it would be appropriate to pack several (p, say) characters into one machine word. In ALGOL such packing is achieved by combining p consecutive components $x[s], x[s+1], \ldots, x[s+p-1]$ of a pseudostring x to an integer

$$\sum_{k=0}^{p-1} x[s+k] \times 128^{p-1-k}$$

and processing this integer like other integers. The *packing density p* hereby depends upon the computer but is considered to be constant for each computer.

The packing of a given pseudostring into a shorter array and the corresponding unpacking could easily, but not efficiently, be described in terms of ALGOL. In fact, in practice such packing and unpacking should always be done by code procedures (cf. § 47). It is therefore more appropriate if we show how we can operate efficiently on packed data with the aid of two code procedures *pack* and *unpack* defined to have the following properties:

pack (x, k, a) is a procedure statement which stores the integer value x $(0 \leq x < 128)$ as the k-th component of a pseudostring which is packed into an array a.

unpack (k, a) is a function designator which produces the value of the k-th component of a pseudostring which is packed into an array a.

Note, however, that for these procedures k does not count the components of the array a but rather the components of the pseudostring as if it were not packed, e.g. with $p = 5$:

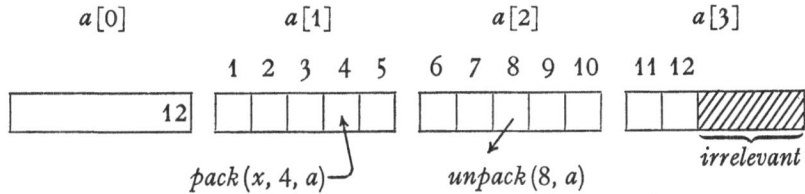

The packing density p is a machine constant and enters only implicitly into these procedures. However, the programmer must also know the value of p for computing the upper subscript bounds for the packed arrays.

With the above convention the concatenation of two pseudostrings (cf. 37.1,a) is now described for packed data as:

```
«begin
    if a[0]+b[0]> bigm then goto overfl ;
    for k := 1 step 1 until b[0] do pack (unpack (k, b), k+a[0], a) ;
    a[0] := a[0]+b[0]
end».
```

Likewise the program for format handling given in 37.2 can be adapted to packed data, as follows:

```
«begin
    for k := 1 step 1 until f[0] do
    begin
        fk := unpack (k, f) ;
        if fk = 10 then x := x/10 ;
        if fk = 12 then goto out
    end ;
out: g[0] := f[0] ;
    sig := (x ≥ 0) ;
    x := abs (x) ;
    if x ≥ 1 then goto overfl ;
    for k := 1 step 1 until f[0] do
    begin
        fk := unpack (k, f) ;
        if fk = 14 then pack (if sig then 42 else 14, k, g)
        else
            if fk = 13 then pack (if sig then 13 else 14, k, g)
            else
                if fk = 10 then
                begin
                    x := 10 × x ;
                    pack (entier (x) + 1, k, g) ;
                    x := x − entier (x)
                end
                else
                    pack (fk, k, g)
    end k
end».
```

The transcription of a program for handling unpacked pseudostrings into one for handling packed data seems obvious. However, if we try to transcribe the sorting program of 37.3, we run into serious difficulties since in this program a two-dimensional array a occurs, every row of which is a pseudostring representing a string of characters, whereas the code procedures *pack* and *unpack* apply only to one-dimensional arrays.

To write a corresponding program for sorting packed data, we need therefore two other procedures which operate upon two-dimensional arrays. The corresponding procedure calls are then

$pack2(x, k, j, a)$ as the equivalent of $a[k, j] := x$

$unpac2(k, j, a)$ as the equivalent of the value $a[k, j]$.

With that, statement ww of the sorting program becomes

«**for** $j := s$ **step** 1 **until** t **do**
  if $phi(ak[j]) \neq phi(unpac2(v[l], j, a))$ **then**
  begin
    $first := phi(ak[j]) < phi(unpac2(v[l], j, a))$;
    goto yy
  end»,

provided we have unpacked the keyword of the k-th data set at the beginning of the k-loop and have stored it as the components s through t of an array ak (this is done for reasons of economy):

«**for** $j := s$ **step** 1 **until** t **do** $ak[j] := unpac2(k, j, a)$ ».

Chapter VI

Declarations

If we are going to write a *complete* ALGOL program, that is, one to be used for actual calculation and not just for explanatory purposes, the following requirements must be carefully observed:

First, the program must have the syntactic form of a *compound statement* or *block,* and

Second, every quantity used in the program must be *declared.*

The first of these requirements is easily met by enclosing the entire program by «**begin**» and «**end**», but in order to meet the second requirement, every quantity to be used in a program must be quoted in an appropriate declaration within that program. However, this otherwise strict rule has three well-defined exceptions:

a) *labels* "declare themselves"; they therefore need no explicit declarations.

b) The *standard functions* and *standard I/O-procedures* (for the latter see Chapter VIII) are considered as permanently declared and therefore need not be declared again in an ALGOL program.

c) *Formal parameters* of a procedure represent no true quantities and therefore need not be declared (but they must be specified; for this see 44.4.2).

The natural question as to *how* a quantity is declared will be answered in § 38— § 41, where the various kinds of declarations are treated, but without considering their environment. The question as to *where* declarations should be placed is answered as follows: Declarations can be given after every «**begin**», e.g.

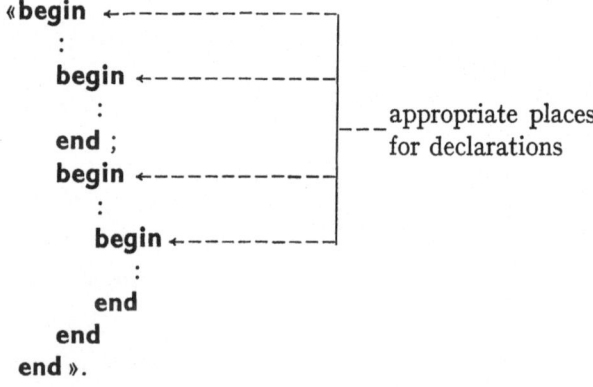

However, by placing a declaration after the «**begin**» of a compound statement the latter is turned into a *block*, a fact which may have severe consequences for the execution of a program. These consequences are treated in § 42, where also the actual meaning of declarations and their interactions with the surrounding program are discussed.

It might be asked why we do not just place all declarations immediately after the first «**begin**» of the program (which indeed is often the most convenient place). The main reason is that the proper distribution of declarations in an ALGOL program is an important instrument for the economisation of storage requirements (cf. 43.3).

§ 38. Type Declarations

Simple variables are declared by type declarations which at the same time define their type (**real**, **integer**, **Boolean**), that is, the type of values that can be assigned to these variables.

38.1. Examples

«**real** *x, y, a12, beta, td/32*»,
«**integer** *k*»,
«**Boolean** *verify, discr, critic*».

These examples state that the names following the declarators «**real**», «**integer**», «**Boolean**» represent simple variables of the respective types.

38.2. Syntax

A type declaration has the syntactic form

$$«T\,I, I, \ldots, I»,$$

where T denotes one of the three symbols «**real**», «**integer**», «**Boolean**», and the I's are arbitrary identifiers denoting simple variables. Syntactic diagram:

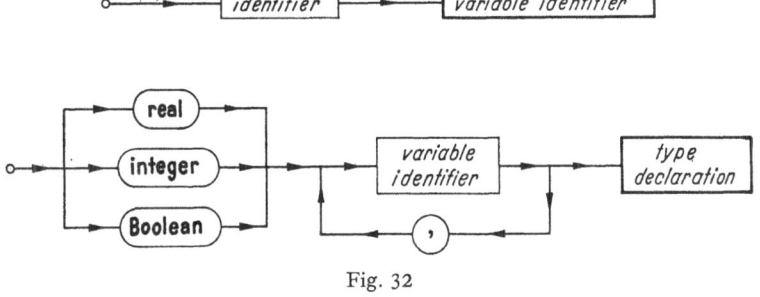

Fig. 32

38.3. Semantics

A type declaration «$T\ I_1, I_2, \ldots, I_k$» declares the simple variables $I_1, I_2, \ldots, I_k$ of type T. As already indicated, such a declaration is only valid for the block at the beginning of which it appears. For further restrictions see § 42.

§ 39. Array Declarations

We recall that an array is a multidimensional arrangement of elements, every one of which behaves like a simple variable, and that these components are distinguished by a set of p integers (subscripts) $i_1, i_2, \ldots, i_p$, where p is called the *dimension* of the array.

An array is declared by an *array declaration* which also defines its dimension p as well as its *extension*, i.e. the array bounds (or subscript bounds) $l_1, \ldots, l_p, u_1, \ldots, u_p$ which delimit the p subscripts of the corresponding subscripted variables.

39.1. Examples

39.1.1. «**array** $a, b, c\,[7:n, -1:i+j]$, $x, y\,[1:10,\ 1:10,\ 1:abs\,(p-3)]$» declares three two-dimensional arrays a, b, c and two three-dimensional arrays x, y, all five being of real type (i.e. all their components are of real type).

39.1.2. «**real array** $q\,[1:n], e\,[0:n]$» declares two one-dimensional arrays[1] (vectors) q, e, both of real type. Note that the leading symbol «**real**» is not actually needed.

39.1.3. «**integer array** $a\,[-1:1,\ 7:8,\ 1:4]$» declares one three-dimensional array a of integer type with explicitly given array bounds. According to the latter, the array consists of 24 components

$$a[-1, 7, 1],\ a[-1, 7, 2],\ a[-1, 7, 3],\ a[-1, 7, 4],\ a[-1, 8, 1],$$
$$a[-1, 8, 2],\ a[-1, 8, 3],\ a[-1, 8, 4],\ a[0, 7, 1],\ a[0, 7, 2], \ldots$$
$$\ldots, a[1, 8, 3],\ a[1, 8, 4].$$

This ordering shows one possible method ("storing by rows") for storing the components of a multidimensional array as a linear sequence.

39.1.4. «**Boolean array** $xxx\,[\textbf{if}\ c<0\ \textbf{then}\ 2\ \textbf{else}\ 1:20]$» declares a one-dimensional array, each of whose components can assume one of the logical values **true** or **false**.

In these examples the array bounds (boundaries of the hyperbox in the index-space) are given partially as explicit numbers, partially as values of variables. For the array xxx, the lower bound is given as a conditional arithmetic expression.

---

[1] Note that in ALGOL terminology a vector is always a one-dimensional array, and that what elsewhere is called the dimension of a vector, is in ALGOL expressed by the array bounds.

39.2. Syntax

An array declaration contains one or more *array segments*, every one of which declares a group of arrays of equal dimension and extension.

39.2.1. An *array segment* has the syntactic form

$$\text{«}I, I, \ldots, I[L:U, L:U, \ldots, L:U]\text{»},$$

where the I's denote arbitrary identifiers (the names of the declared arrays) and the L's and U's are subscript expressions. The syntactic construction «$L:U$» is called a *bound pair* and defines the subscript bounds for the corresponding subscript position.

39.2.2. An *array declaration* has the syntactic form

$$\text{«}T \text{ array } G, G, \ldots, G\text{»},$$

where T represents either blank space or one of the declarators «**real**», «**integer**», «**Boolean**», and every G stands for an array segment.

39.2.3. *Syntactic diagram*:

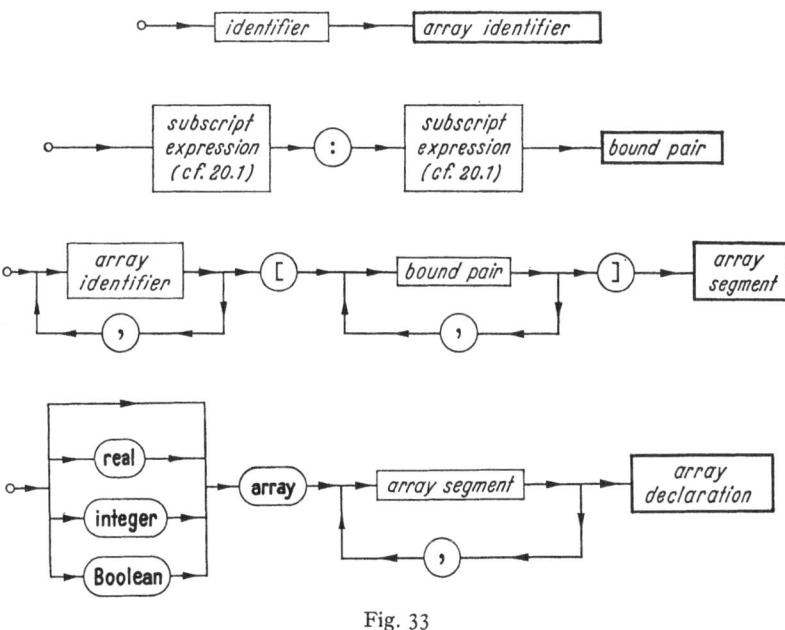

Fig. 33

39.3. Semantics

39.3.1. Every array segment

$$\text{«}I_1, I_2, \ldots, I_q[L_1:U_1, L_2:U_2, \ldots, L_p:U_p]\text{»}$$

of an array declaration declares the q arrays $I_1, I_2, \ldots, I_q$, all of which
have the same dimension p and the same array bounds; the latter are
defined by the actual values of the subscript expressions $L_1, L_2, \ldots, L_p$,
$U_1, U_2, \ldots, U_p$.

In other words, every one of the arrays $I_1, I_2, \ldots, I_p$ consists of
exactly those components $I_j[K_1, K_2, \ldots, K_p]$ for which the subscripts K_i
satisfy the condition

$$\text{``value of } L_i\text{''} \leqq K_i \leqq \text{``value of } U_i\text{''} \quad (i = 1, 2, \ldots, p),$$

the values being taken at the moment of entry into the block in which
the array declaration appears.

39.3.2. The whole array declaration

$$\text{« } T \text{ array } G_1, G_2, \ldots, G_m \text{ »}$$

declares the total of all arrays appearing in all array segments $G_1, G_2,$
$\ldots, G_m$. All these arrays are of the same type defined by the declarator T.
If T is blank space, then the arrays are of type **real**; therefore both
declarations

$$\text{« array } a[1:n] \text{ »} \quad \text{and} \quad \text{« real array } a[1:n] \text{ »}$$

have the same meaning.

39.3.3. *Evaluation of array bounds.* The subscript expressions L, U occurring in an array declaration are evaluated when the declaration is
encountered (i.e. upon entry into the block), and the actual values of
all variables involved in these expressions are taken at that moment.

The array bounds thus evaluated remain valid until the exit from
the block, even if the variables from which they were computed change
their values. As an example, the array bounds of the array a in the
following piece of program remain $[1:7]$ throughout the execution of
block b:

```
    «   n := 7 ;
b:  begin
        array a[1:n] ;
        n := 2 × n ;
            ⋮
    end ;».
```

39.3.4. *Empty arrays.* If in an array segment one of the array bounds U_k
yields a value below the corresponding lower bound L_k, then all arrays
declared by that array segment are empty (i.e. they have no components),
and any reference to a component of one of these arrays will be undefined.

Accordingly «**array** $a[1:n]$; $a[1] := 0$» is an *incorrect* piece of ALGOL program if $n < 1$, whereas

    «**begin**
        array $a[1:n]$;
        for $k := 1$ **step** 1 **until** n **do** $a[k] := 1$; ...»

is allowed even if $n < 1$, since in that case the reference to the array a is suppressed by the rule for empty for-list elements (cf. 30.5.2).

39.3.5. *Undefined array bounds.* In order that none of the subscript expressions L, U occurring in an array declaration yield an undefined value (in which case the program would be defective), all variables appearing in bound pairs must have defined values when the block is entered. This in turn implies the following rule:

All variables and other quantities occurring in bound pairs of an array declaration must be declared *outside* the block in whose head the array declaration is located.

The following is therefore not allowed:

    «**begin**
        integer ji ;
        array $wrong[1:ji]$; ...».

For an array declaration given at the very beginning of a program, the above rule has the consequence that the subscript expressions L, U can contain only explicitly given values, e.g.

    «**array** $a[1:17, -1:3]$, b, $c[1:entier(exp(5))]$,
        $d, e, f[length('abc \sqcup de/f'):entier(2\uparrow7.251)]$».

39.4. Unused components of an array

There is of course no obligation to use all the components of an array. Therefore, if for a certain computation approximately the first 50 coefficients of a power series will be used, but nothing more is known about their exact number, we may begin the program with

    «**begin**
        real array $coeff[0:200]$; ...».

The possibly unused components $coeff[51]$ through $coeff[200]$ will do no harm, whereas a too small upper bound will certainly cause trouble.

Another example are symmetric matrices. For these, many numerical methods can be programmed such that the subdiagonal elements do not enter the computation, a fact which reduces computing time considerably. However, because of the syntax and semantics of array declarations, always the full matrix must be declared (e.g. «**array** $a[1:n, 1:n]$»), while the subdiagonal elements are simply not used.

On the other hand, one should not be too generous with respect to array bounds because the translated program has to provide storage space for all components of a declared array whether they are used or not.

§ 40. Switch Declarations

A switch declaration declares a certain identifier to represent a *switch*, i.e. a selecting mechanism which serves to give corresponding switch designators (cf. 25.2) a meaning.

40.1. Examples

«**switch** *wernik* : = *arica, acryl, m17, larix*»,

«**switch** *llll* : = *nora*».

Switch *wernik* allows selecting one of the four labels listed, e.g. «*wernik*[3]» will select label *m17*. In the second example the switch list contains only one label, which is allowed but not very meaningful (it allows only the application «**goto** *llll*[1]»).

40.2. Syntax

A switch declaration has the syntactic form

«**switch** I : = $L, L, \ldots, L$»,

where I is an identifier (the switch identifier, i.e. the name of the switch) and the L's denote arbitrary (source-) labels. The syntactic construction «$L, L, \ldots, L$» is called the *switch list*.

Syntactic diagram:

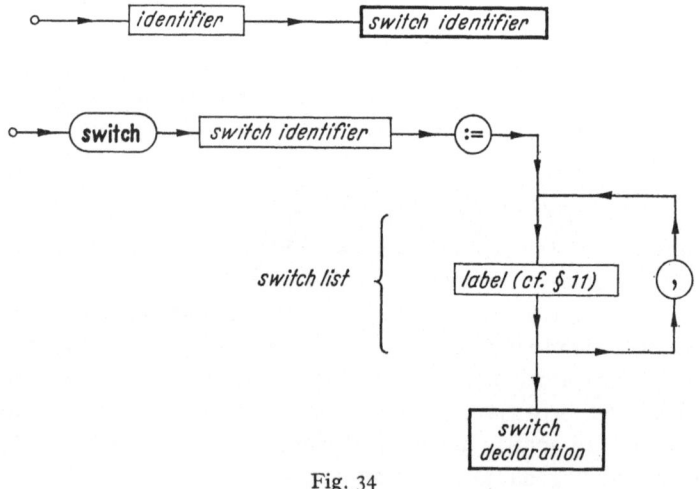

Fig. 34

40.3. Semantics

The selecting mechanism of a switch I declared by

$$\text{«switch } I := L_1, L_2, \ldots, L_p\text{»}$$

works as follows: If a switch designator «$I[E]$»[1] (where I is the same identifier as above) is encountered, the following actions take place:

a) The expression E is evaluated and its value is rounded to the nearest integer.

b) If the value thus obtained is k, and if $1 \leq k \leq p$, then the source-label L_k is selected from the switch list and used in place of $I[E]$.

c) If the value k is <1 or $>p$, (where p is the number of entries to the switch list) then the switch designator $I[E]$ is undefined.

Accordingly, «**goto** *wernik*[2]», where switch *wernik* is defined as in 40.1, is equivalent to «**goto** *acryl*», but as a consequence of c) such statements as

«**goto** *ammon*[0]»,   or

«**goto** *wernik*[k]»,   where e.g. $k = -3$ or $k = 5$,

are undefined.

40.4. Influence of scopes (compare also § 42)

40.4.1. A switch designator «$I[E]$» must of course be within the scope of switch I and also within the scopes of all quantities which appear in the expression E. However, the destination of a jump «**goto** $I[E]$» may well be outside the scope of I.

40.4.2. *The environment rule for switches.* It is an absolute must that every one of the source labels L_k appearing as entries of a switch list be within the scope of a matching destination label L_k. It is always this latter which is considered as the destination label *corresponding* to the source label L_k.

In other words, *while the expression E in a switch designator «$I[E]$» is evaluated at the location of the latter, the destination of the jump is selected at the location of the declaration of switch I.*

40.4.3. *Example.*

```
→ «aa:  begin
            switch abc := aa, bb, cc ;
→    bb:     begin
                integer aa ;
     bb:         aa := k ;
     cc:         goto abc[aa]
             end ;
→    cc: end».
```

---

[1] Note that in SUBSET ALGOL 60 switch designators can occur only after the symbol «**goto**».

Obviously the switch designator in «**goto** $abc[aa]$» meets the requirements of 40.4.1, and also 40.4.2 is fulfilled since for every one of the three entries in the switch list there is indeed a corresponding destination label (these are indicated by arrows). Thus the jump «**goto** $abc[aa]$» will, if e.g. $aa = 1$, be directed to the destination label aa at the beginning of this program fragment. It is no offense against the rules that this label aa is suppressed at the location of the jump, nor that it is outside the scopes of both switch abc and integer variable aa.

If on the other hand $aa = 2$, then the jump will not be directed to the label bb in front of the statement «$aa := k$» but to the label bb in front of the inner block since the switch declaration is within the scope of the latter. It is interesting to note that within block bb «**goto** $abc[2]$» is *not* equivalent to «**goto** bb».

40.4.4. A *counterexample*:

```
«begin
    switch bmn := x, y, z ;
aa: begin
        integer kappa ;
    x:      ⋮
    y:      ⋮
    z:      ⋮
            goto bmn[kappa]
        end
    end».
```

The jump «**goto** $bmn[kappa]$» is correctly within the scopes of both bmn and $kappa$; however, the destination labels x, y, z are valid only within block aa. Hence, if no other labels x, y, z which meet the conditions stated in 40.4.2 exist outside this program fragment, the declaration for switch bmn is illegal. The situation might be corrected by inserting the switch declaration immediately after the declaration for $kappa$.

§ 41. Procedure Declarations I[1]

A procedure declaration defines a *procedure*, which is a prefabricated Algol subroutine that may later be called through a corresponding procedure statement or function designator. Besides stating the identifier to be used as name of the procedure, the procedure declaration contains a list of the *formal parameters* which denote the formal operands of the procedure, a *specification part* defining kinds and types of all formal operands, and finally the *procedure body* which is a description of the procedure in terms of Algol or as a piece of code.

---

[1] In this section we describe mainly the syntax of procedure declarations. For the semantics see Chapter VII.

Procedures fall into two subclasses: The *ordinary procedures*, which are called by procedure statements, and *function procedures*, which are called through function designators.

41.1. Examples

41.1.1. Declaration of an ordinary procedure which is modelled after the bisection routine given in 30.5.5:

«**procedure** *bisect* (*eps*, *f*, *a*, *b*) ; $\begin{cases} \text{Procedure identifier,} \\ \text{formal parameter part.} \end{cases}$

    real *a*, *b*, *eps* ; **real procedure** *f* ;   Specification part.

    begin

        real *x* ;

        for $x := (a+b)/2$

            while $b-a > eps \wedge x \neq a \wedge x \neq b$ **do**   Procedure body.

            if $f(x) < 0$ **then** $a := x$ **else** $b := x$

    end *bisect*».

41.1.2. Ordinary procedure without formal parameters:

«**procedure** *max* ;   Procedure identifier.

    if $x < y$ **then**

    begin

        real *z* ;   Procedure body.

        $z := y$; $y := x$; $x := z$

    end».

41.1.3. Function procedure which computes the distance of a given point *x*, *y* from the origin:

«**real procedure** *rad* (*x*, *y*) ; $\begin{cases} \text{Procedure identifier,} \\ \text{formal parameter} \\ \text{part.} \end{cases}$

    real *x*, *y* ;   Specification part.

    $rad := sqrt(x\uparrow 2 + y\uparrow 2)$ ».   Procedure body.

41.2. Syntax

41.2.1. The declaration of an *ordinary procedure* has one of the following *basic* syntactic forms (for *structurized* forms cf. 41.2.3 and 44.4.3):

        «**procedure** *I* ; *S*»,

        «**procedure** *I*(*F*, *F*, ..., *F*) ; *V C S*»,

with the following constituents:

I   denotes the *procedure identifier*, i.e. the name of the procedure to be declared.

$(F, F, \ldots, F)$ is the *formal parameter part*. The F's are the *formal parameters*, i. e. identifiers denoting the formal operands of the procedure (according to SR, item 5.4.3, the F's must be different from each other).

V represents the *value part*, which is either empty or has the syntactic form

$$\text{«value } F, F, \ldots, F \text{ ;»},$$

where the F's are identifiers selected from the formal parameter part.

C represents the *specification part*, which is either empty or consists of a sequence of specifications (in juxtaposition), each of which has the form

$$\text{«} Z F, F, \ldots, F \text{ ;»}.$$

Here the F's are again identifiers selected from the list of formal parameters[1], and Z denotes one of the 14 possible *specifiers*:

«real», «integer», «Boolean», «array», «real array», «integer array», «Boolean array», «label», «switch», «procedure», «real procedure», «integer procedure», «Boolean procedure», «string».

S stands for the *procedure body*, which may be an arbitrary statement or a piece of code (for the latter see 47.3).

The construction «I $(F, F, \ldots, F)$; VC» is sometimes called the *procedure heading*.

41.2.2. For a *function procedure* the syntactic form of the declaration is the same, except for the following deviations:

a) An additional declarator **«real», «integer»** or **«Boolean»** must be placed in front of the procedure declaration, which must thus begin with **«real procedure ...»** or **«integer procedure ...»** or **«Boolean procedure ...»**.

b) The procedure identifier must occur within the procedure body as assignment variable in at least one assignment statement just as if it were a simple variable, but it may not occur otherwise in the procedure body.

41.2.3. *Structurized forms of the formal parameter part*[2]. It is allowed to replace any of the commas separating the formal parameters in the formal

---

[1] According to SR, item 5.4.5, every formal parameter F of the procedure must be quoted exactly once in the specification part.

[2] For the special structurisation recommended for procedures to be published in this Handbook, see 44.4.3.

parameter part by a syntactic object

$$\langle\!\rangle\, XX \ldots X \colon (\!\rangle,$$

where the X's represent arbitrary letters. This optional device allows grouping the formal parameters into categories. As an example, the declaration for procedure *bisect* (41.1.1) could be modified into

«**procedure** *bisect* (*eps, f*) *trans*: (*a, b*) ; …».

Such a modification does not influence the properties of the procedure, provided the order in which the formal parameters are listed is maintained.

41.2.4. *Syntactic diagram*:

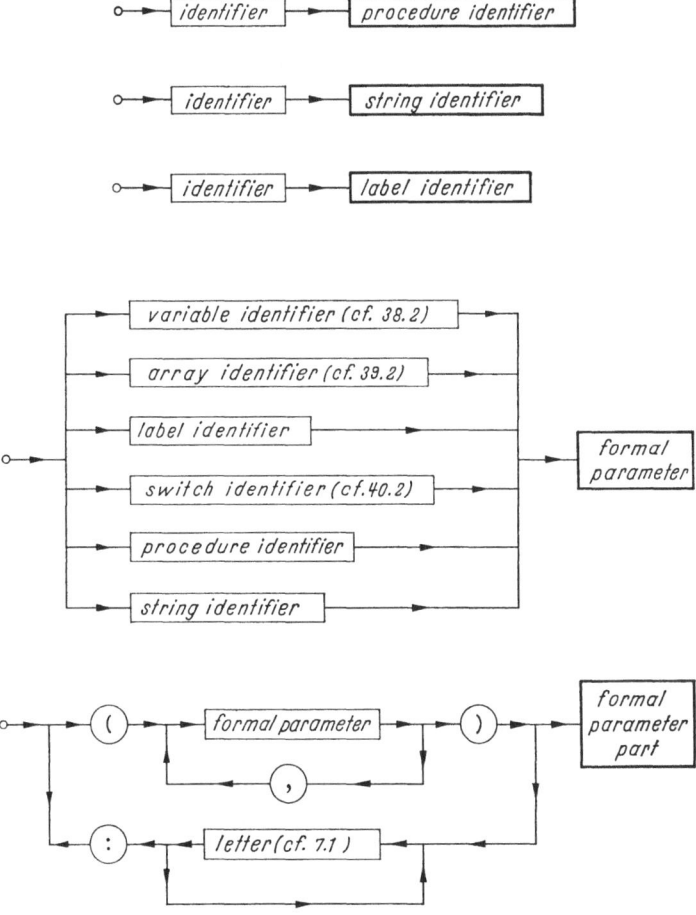

Fig. 35 a

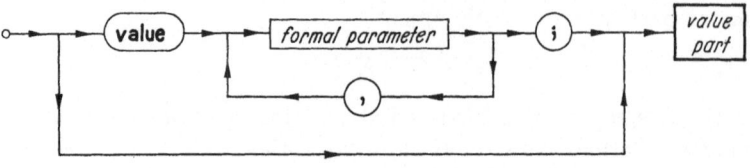

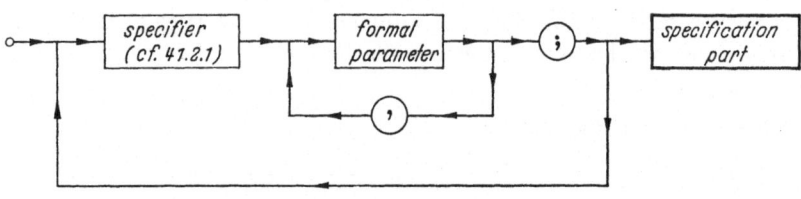

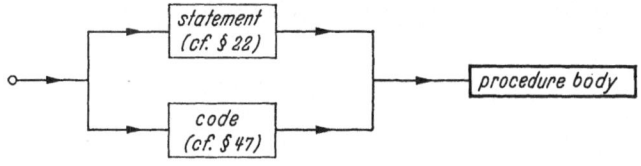

Fig. 35 b

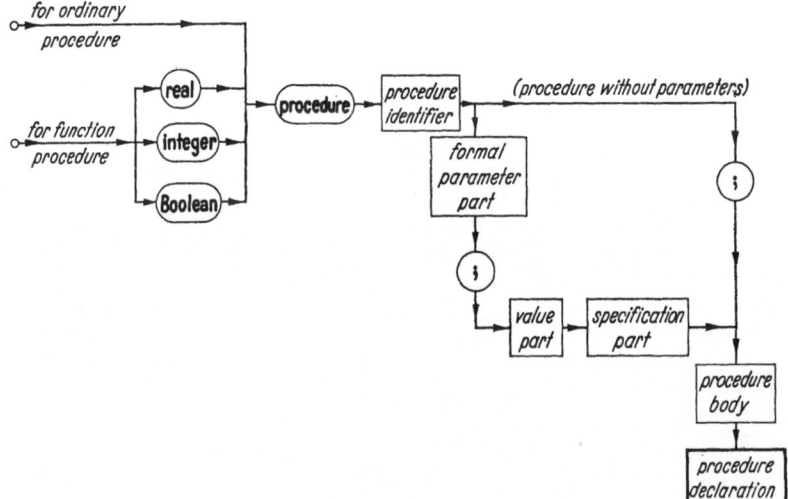

Fig. 35 c

41.3. Semantics

The semantics of procedure declarations will be treated in great detail in § 44. Here we mention only that a procedure declaration serves to give corresponding procedure calls (function designators, procedure statements) a meaning, but itself never causes the execution of any operations when it is encountered during execution of a program. In addition, it is impossible to jump from outside into a procedure declaration (even if the procedure body is not a block).

Furthermore, the formal parameters of a procedure declaration do not represent true quantities as identifiers usually do, but are merely designations given to formal objects (the formal operands of the procedure) which have a meaning only inside the procedure declaration.

§ 42. Semantics of Blocks

At the beginning of a block, one or more declarations may be given, e.g.

```
«begin
    real t ;
    integer j ;
    integer array b[0:n] ;
start:
    b[0] := entier(x) ;
    t := x − b[0] ;
    for j := 1 step 1 until n do
    begin
        t := 1/t ;
        b[j] := entier(t) ;
        t := t − b[j]
    end ;
comment Output of partial denominators of continued fraction ap-
        proximation for the given real number x using the standard
        I/O procedure outarray (cf. Chapter VIII) ;
    outarray(1, b)
end of block».
```

These declarations declare certain identifiers (here t, j, b) to represent certain quantities within that block. Furthermore, all destination labels in the block (here only *start*) also act like declarations which declare the label identifiers as names of spots in the program. All these quantities exist only within the block, whereas the other quantities occurring in that block (in the above example x, n, *entier*, *outarray*) exist also outside.

42.1. Block structure

A block in an ALGOL program may contain other blocks which in turn may contain further blocks in their interior, and so on. Thus a block B has a structure which may be visualized by a diagram in which ordinary ALGOL text is represented by horizontal lines, but the **begin**'s and **end**'s of the various blocks (but not those of ordinary compound statements!) are indicated by a step upwards, respectively downwards:

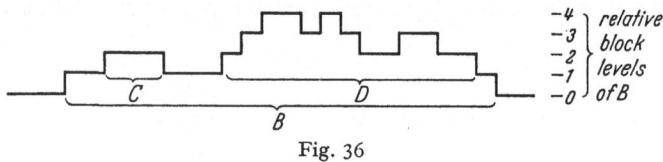

Fig. 36

The various levels in the above diagram correspond to what we call the *block levels*: The level 0 is called the *environment* of block B; it is that block level into which B is embedded. Level 1 is the *block floor* of B; it is the lowest block level inside B and contains the declarations for the block. The terms *environment* and *block floor* are to be understood relative to the block to which they apply; for instance, the floor of block B is at the same time the environment for the subblocks C and D, while the environment of B is the floor of the smallest block containing B.

These terms apply also to the whole program (provided it is a block); we then speak of the *program floor*, block levels of the program, etc., and these are indeed absolute terms.

42.2. Scope of a quantity[1]

42.2.1. Every quantity occurring in an ALGOL program has a *scope*, that is, a region (of the program) in which the quantity exists and can be referred to by its identifier. Outside the scope the quantity is non-existent and cannot be used; its identifier is either meaningless or refers to another quantity.

42.2.2. The concepts *local, global* and *suppressed*. If an identifier I appears either as destination label in the floor of block B or is declared[2] there,

---

[1] The rules given here apply so far only outside procedure declarations; in § 44 we shall give additional rules which cover the most general case.

[2] To be precise, an identifier is "declared" in the floor of a block B if it appears in a declaration in that block floor in one of the positions which are denoted by I in the syntactic forms given in 38.2, 39.2.1, 40.2, 41.2.1.

it represents a quantity Q which exists only inside B and is called *local* to B. Usually the scope of such a quantity Q is the full block B, as is the case for the quantities t, j, b declared in the above example. However, if B has subblocks, the following is more precise:

The quantity Q exists *at least* in the floor of the block B to which it is local. For the higher block levels we can decide the existence of Q by recursive application of the following rule: If Q exists in the environment of a subblock Bx of B, then either

a) The identifier I of Q is not declared or used as destination label in the floor of Bx; in that case Q is called *global* to Bx, which is to say that it exists also in the floor of Bx.

b) The identifier I of Q is declared or used as destination label in the floor of Bx; then Q is called *suppressed* in Bx, which means that it is nonexistent throughout Bx because the identifier I is used there as name of another quantity Q^*.

The following diagram exhibits the various possibilities:

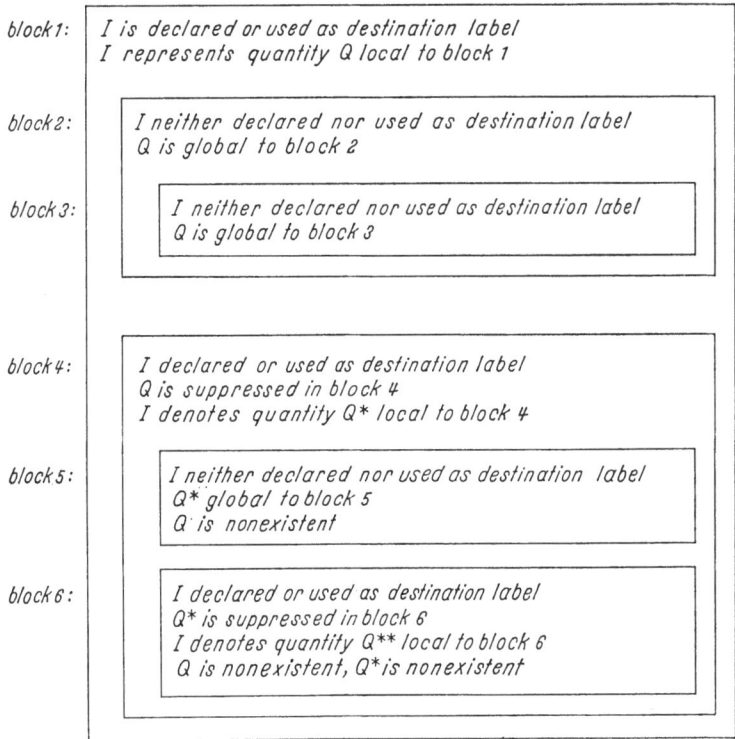

Fig. 37

42.2.3. To sum up, *the scope of a quantity Q* is exactly the block B to which it is local, minus all subblocks in which the identifier I of Q is declared again or appears as destination label, and — for the present — also minus the interiors of all procedure declarations contained in B (see however 44.5.1).

In the following (utterly unrealistic) piece of program the scopes of the various quantities

1) **label** d ; **integer** m ;
2) **real** x, z ; **label** al ;
3) **real** y ; **label** e ; **integer** n ;
4) **procedure** x ;
5) **integer array** al ; **label** f, z ;
6) **switch** x ;

are indicated on the right side (+ denotes existing quantities, — suppressed quantities and ○ otherwise nonexistent quantities):

```
                                                        1 2 3 4 5 6
«d: begin
al:      real x, y, z ;  integer n ;                    + + + ○ ○ ○
         x := 1 ; y := 2.0↑m ; n := m↑2 ;               + + + ○ ○ ○
e:       begin
             integer array al[1:2×n] ;                  + − + + + ○
             procedure x (n) ;
                 value n ;  integer n ;
                 begin
                     integer z ;
                     for z := 1 step 1 until n do
                         al[z] := z
                 end x ;
z:       if y > n then goto e ;                          + − + + + ○
         x (n − 1) ;                                     + − + + + ○
f:           begin
                 switch x := d, e, z, f ;                + − + − + +
                 y := al[n + 1] := al[n − 1] ;           + − + − + +
                 if al[1] > y then goto x[m]             + − + − + +
             end f ;
             y := al[n] ;                                + − + + + ○
         end e ;
         m := entier (ln (y))                            + + + ○ ○ ○
     end d».
```

For scopes see 44.5

42.2.4. *Remark*: According to what has been said above, it may happen (and is allowed) that an identifier I of a quantity local to a block B is declared again or used as destination label in the floor of a subblock Bx of B. However, this should not be considered as a regular programming practice, though it cannot always be avoided without undue effort.

42.2.5. *Scopes of standard-functions and -procedures.* The 10 standard functions and the 6 standard I/O-procedures are also regular quantities (procedures) having scopes like any other quantity. As a rule, the scope of these standard quantities encompasses all ALGOL programs, but again with the possible exception of certain subblocks.

An immediate consequence of this is that it is allowed (but not recommended) to write a block as follows:

```
«begin
    real array sin [0:100] ;
    switch inreal := take, put, store, get ;
exp: ... end».
```

However, if this is done, the standard functions *sin, exp* and the standard I/O-procedure *inreal* are suppressed in this block and cannot be used there; indeed, a function designator «*sin (z)*» occurring in this block would be undefined.

42.3. Restrictions for declarations

42.3.1. Whereas we have the freedom to declare the same name several times in one program, we cannot declare it several times in the same block floor. The latter is forbidden by the following rule:

No identifier may be declared more than once in the same block floor, nor be declared and at the same time be used as destination label in a block floor, nor be used more than once as destination label in the same block floor.

In the example below three violations of this rule occur, one of them being due to the rule given in 9.2.4, which says that e.g. *label33* and *label34* are not distinguished in the subset. On the other hand, no conflict arises because the identifier c is declared and used as formal parameter in the same block head:

```
«begin
    real a, beta, c ;
    procedure neg (c) ;  integer c ;  c := − c ;
    switch a := pale, sleek, quick, sleepy ;
label33: ... ;
label34: ... ;
beta:    ... ;
    end».
```

42.3.2. A further rule (see SR, item 5) says:

Except for labels, a quantity Q cannot be used[1] between the «**begin**» of the block to which Q is local and the semicolon that terminates the declaration for Q.

This rule is of importance only in connection with procedures and is therefore exemplified in Chapter VII; in fact, the only other examples to which it would seem to apply are of the type

> «**begin**
>   **array** $x[1:n]$;
>   **integer** n ;
>   ...»,

but this is already forbidden by 39.3.5.

42.4. Dynamic effects of declarations

42.4.1. The only declarations which, upon execution of the program, cause the execution of operations on the ALGOL level are the array declarations; indeed, for these the array bounds must be obtained by evaluation of the subscript expressions in the bound pairs at the moment of entry into the block. These array bounds then remain unchanged during the whole execution of that block, even if the variables used to calculate them change their values.

42.4.2. Except for the evaluation of array bounds, the declarations at the beginning of a block B do not incite any actions on the ALGOL level upon entry into the block but solely state that in block B certain identifiers will denote new quantities, while the external quantities (if any) which used the same names outside B become suppressed. On the other hand, the declarations at the beginning of a block will initiate certain operations on the object program level; for instance, the storage reservations for all quantities declared in that block will be made upon entry into the block.

After these preparations the execution of the statements of the block begins; it should be recognized, however, that at that moment the values of all local simple variables and local arrays of the block are still undefined. This has the consequence that a piece of program like

> «... ; $m := 2$;
> $k := k + 1$;

---

[1] Use of a quantity Q means any occurrence of its identifier, with the exception of the position in which the identifier is declared (see footnote 2 on p. 158). Furthermore, the occurrence of a function procedure identifier as assignment variable in the procedure body, as required by the rules given in 41.2.2.b, is not a "use" of the function procedure but a constituent of its declaration.

begin
   integer k ;
   $p := k\uparrow 2 + m\uparrow 2$;
   ... »

is always meaningless because the k involved in the expression $k\uparrow 2 + m\uparrow 2$ is still undefined when it comes time to evaluate the expression; in fact, this k is a new variable entirely different from the k occurring outside the block.

42.4.3. *Exit from a block* occurs as soon as the execution of the statements in the block proceeds to the «**end**» of the block or if a goto-statement is encountered which points to a destination label outside the block (including a destination label in front of the block). In either case all quantities local to the block lose their existence, and their values (if they had any) are destroyed[1]. What actually occurs on the object program level at that moment is that all storage reservations for the block are cancelled again; the storage space thus set free is used automatically for subsequent storage reservations.

At the same time the quantities which were suppressed in that block come into existence again and resume the values they had at the time of entry into the block[2, 3], e.g.:

«
    ⋮
$k := k + 1$;   { "outer" variable k assumes a value X.
    begin
        integer k ;   "outer" variable k suppressed, k represents a new variable called the "inner" variable k.
        $k := m\uparrow 2 - 1$;
        ⋮
    end ;
$p := k\uparrow 2 + m\uparrow 2$;   "inner" variable k undefined, "outer" variable k has again value X, which is used in this expression.
    ⋮
    ».

---

[1] In most compilers this destruction is not actually performed upon exit from a block but will eventually take place through an assignment statement which overwrites the value. It may thus occur that the corresponding object program still has access to such values after exit from the block; as a consequence, correct results may be produced by an incorrect ALGOL program.

[2] Under certain circumstances (cf. 45.2.4) the value of a suppressed quantity may change during its suppression.

[3] The reader should be aware of the fundamental difference between exit from a block to which a variable V is local and entry into a subblock in which V is suppressed. In the second case the value of V is not destroyed, and the storage reservations for V are not cancelled.

42.4.4. In case a block is left and re-entered several times, the storage reservations are made anew for every entry into the block and are cancelled again upon every exit. Thus in a case like

«**for** $k := 1$ **step** 1 **until** n **do**
   begin
      array $c[0:k]$;
        $\vdots$
   end»,

the declarations given at the beginning of the block are activated anew for every repetition of the controlled statement, but the declared quantities lose their existence and their values again every time the «**end**» of the block is met. Accordingly, in the above example the values of the components of the array c cannot be saved from one execution of the controlled statement over into the next (see, however, footnote 1 on p. 163).

42.5. Operands of a block

42.5.1. The operands of a block B are defined as those quantities existing outside the block which are involved in the execution of the block. Obviously the quantities global to B are operands of B provided *they are actually used inside B.*

In the example

«**begin**
    real x, y ;
     $\vdots$
z: **begin**
       real t ;
       $t := x$;
       $x := sqrt(t)$;
     end z ;
     $\vdots$
   end»

$x, y, sqrt$ and the label z (as far as we can see) are global to block z, but only x and $sqrt$ are operands since y and z are not used in block z, which on the whole is equivalent to the operation «$x := sqrt(x)$».

42.5.2. *Hidden operands.* A block may have hidden operands: Indeed, if a procedure P is operand of block B, then — according to the environment rule for global parameters (cf. 44.3.2) — the global parameters of P are also involved in the execution of B, hence operands of B. We call these hidden operands of B because they cannot be found by inspection of block B but only by inspection of the declaration for procedure P, which is given somewhere outside B.

Consider for instance the following piece of program:

```
«begin
    real x ;
    procedure nix ;                                      ⎫ Declaration of
        begin x := 0 ;  goto out end ;                   ⎭ procedure nix.
             ⋮
abc: begin
        integer x, out ;
             ⋮
        nix                                              ⎧ Call of procedure
    end abc ;                                            ⎩ nix.
out:
    end».
```

Here the real type variable x and the label *out* are — although suppressed in block *abc* — (hidden) operands of block *abc* since procedure *nix* is called in that block, and x as well as *out* are global parameters of procedure *nix*.

42.5.3. The operands of a block B fall into the following four categories:

a) *Arguments*: Single variables and arrays, the values of which influence the execution of the block without being changed themselves, and procedures (because these also influence the execution of the block).

b) *Results*: Simple variables and arrays to which during execution of the block new values are assigned without making use of the values they had upon entry into the block.

c) *Transients*: Simple variables and arrays which have properties of both arguments and results: Their values upon entry to the block influence the execution of the block, but are themselves changed during execution of the block.

d) *Exits*: Labels referring to destinations located outside B and switches which are declared outside B. In both cases the only operation that can be due to such an operand is a jump from inside B to a destination outside B[1], thus terminating the execution of B.

These categories are not distinguished syntactically, nor has the programmer to take measures such as treating quantities of the four categories differently, but this classification is of some importance for discussing the properties of a procedure.

§ 43. Entire Programs

An ALGOL program is a piece of ALGOL text which describes a complete computation including the transfer of initial data from outside

---

[1] For switches this follows immediately from the environment rule for switches (cf. 40.4.2).

into the computer and of the final results from inside the computer to the outside world.

43.1. Rules for ALGOL programs

43.1.1. An ALGOL program must have the syntactic form of a block or compound statement.

43.1.2. With the exception of labels, formal operands, standard functions and standard I/O-procedures, all quantities occurring in an ALGOL program must be properly declared.

43.1.3. For array declarations given at the very beginning of an ALGOL program (in the program floor), the subscript expressions L, U which define the array bounds can only be *explicitly given values*.

43.1.4. If an ALGOL program is not a block, then 43.1.3 must hold for the array declarations given in the floor of the biggest block contained in the program.

43.1.5. All **begin**'s and **end**'s of an ALGOL program must form a complete *begin-end-structure*, i.e. to every «**begin**» there must exist exactly one corresponding «**end**» (and vice-versa) such that the two delimit a compound statement or block. It is advisable to check every ALGOL program carefully for this property, e.g. by scanning the program and stepping a counter one upwards before every «**begin**» and one downwards after every «**end**» (excepting of course **begin**'s and **end**'s in comment situations). The begin-end-structure is correct if and only if the counter comes down to zero after the final «**end**» but never before. This method would readily disclose the following begin-end-structures as being incorrect:

«**begin begin end** ; **begin end end** ; **begin end**»,
«**begin begin end** ; **begin begin end** ; **begin end end**».

43.1.6. The final «**end**» of an ALGOL program has the dynamic effect of a stop[1].

43.2. Examples

For the examples to be given in this section we assume for the moment the existence of procedures *line, print, read, prtext* as if they were standard procedures[2], the calls of which have the following properties:

---

[1] There is no special "stop statement" in ALGOL 60 (in contrast to ALGOL 58). Instead, one places a labelled dummy statement in front of the final «**end**» of the program and jumps to it.

[2] These are not the standard I/O-procedures of ALGOL 60 (for which see Chapter VIII) but are just assumed for the purposes of § 43. They can be expressed, however, with the aid of the true standard I/O-procedures (cf. 49.5).

«line»   causes a line feed of the printer.
«print (E)»   prints the value of the arithmetic expression E with a
                  certain standard format.
«read (V)»   reads the next number from the input medium and assigns
                  it to the real type simple variable V.
«prtext ('...')» prints the text enclosed between the string quotes.

43.2.1. «**begin** *line*; *print (sqrt (3))* **end**» is a complete ALGOL program,
the action of which is the printing of the value $\sqrt{3}$.

43.2.2. The following program prints a function table; for every $x = 0.01$,
$0.02, \ldots, 99.99, 100.00$ the values of x and the values of the functions
sin, cos, sqrt, exp, ln, arctan for this x are printed on one line:

```
«begin
    integer k ;
    for k := 1 step 1 until 10000 do
    begin
        line ;
        print (k/100) ;
        print (sin (k/100)) ;
        print (cos (k/100)) ;
        print (sqrt (k/100)) ;
        print (exp (k/100)) ;
        print (ln (k/100)) ;
        print (arctan (k/100)) ;
    end
end».
```

43.2.3. Solution of the quadratic equation $ax^2 + bx + c = 0$ with consideration of all possible cases:

```
«begin
    real a, b, c, d, e, p, q, x, y ;
    read (a) ; read (b) ; read (c) ;
    line ; line ; line ; line ;
    print (a) ; print (b) ; print (c) ;
    line ;
    if a = 0 then
c1: begin
        if b = 0 ∧ c = 0 then prtext (' arbitrary ⊔ x')
        else
            if b = 0 ∧ c ≠ 0 then prtext (' no ⊔ solution')
            else
```

$c2$:   **begin**
                                prtext ('*single* ␣ *solution* ␣ ␣ ␣ ') ;
                                print ($-c/b$)
                          end $c2$

   end $c1$
   else
$c3$:   **begin**
                $p := b/a$;
                $q := c/a$;
                if $q=0$ **then begin** $e := abs\,(p/2)$; $d := 0$ **end**
                else
                   if $abs\,(p) \geqq 1$ **then**
$c4$:   **begin**
                             $d := 1/4 - q/p/p$;
                             $e := sqrt\,(abs\,(d)) \times abs\,(p)$;
                          end $c4$
                          else
$c5$:   **begin**
                                 $d := q \times (p/q \times p/4 - 1)$;
                                 $e := sqrt\,(abs\,(d))$
                              end $c5$;
                   if $d \geqq 0$ **then**
$c6$:   **begin**
                             $x := -p/2 - ($**if** $p > 0$ **then** e **else** $-e)$;
                             $y := $ **if** $x=0$ **then** 0 **else** q/x ;
                             prtext ('*real* ␣ *solutions* ␣ ␣ ␣ ') ;
                             print (x) ;
                             print (y)
                          end $c6$
                          else
$c7$:   **begin**
                                 $x := -p/2$;
                                 $y := e$;
                                 prtext ('*complex* ␣ *solutions* ␣ ␣ ␣ ') ;
                                 print (x) ;
                                 prtext ('␣ ␣ ␣ $+i$ ␣ ') ;
                                 print (y)
                              end $c7$
             end $c3$;
       line ; *line* ; *line* ; *line* ;
   end ».

No jumps are needed in this program because the if-then-else-structure is such that after printing text and solution (if any) the computation proceeds automatically to the « **end** » of the program where it comes to a stop.

As an example, the output of this program for $a=2$, $b=-5$, $c=2$ would be

2.0000 0000	− 5.0000 0000	2.0000 0000
real solutions	2.0000 0000	$5.0000\,0000_{10}-1$

43.2.4. *Inversion of a matrix of arbitrary order.* Since in this problem a matrix of arbitrary order n will occur, it must be declared as « **array** $a[1:n,\, 1:n]$ », where the value of n will be read from an input medium. However, such a declaration can never appear in the program floor (see 43.1.3, 39.3.5) but only at some higher block level. Thus we shall have at least two block levels.

The method we choose for the inversion is essentially the Gauss-Jordan method [29], but for simplicity we use the diagonal elements as pivot elements, namely $a[p,\, p]$ as pivot in the p-th elimination step. As a consequence, the basic formulae of the process (formulae (5) in [29], corresponding to statements *alpha, beta, gamma, delta* in this program) are used with $p=q$:

```
«begin
     real x ;
     integer n, i, j, p ;
     read (x) ;   n := x ;
     begin
          array a[1:n, 1:n] ;
     comment Follows input of the matrix elements a[i, j] one by one ;
          for i := 1 step 1 until n do
               for j := 1 step 1 until n do
                    begin read (x) ;   a[i, j] := x end ;
     comment Follows inversion of matrix A "on the spot", i.e. the ele-
               ments of the inverted matrix are again stored as a[i, j] ;
          for p := 1 step 1 until n do
loop:     begin
               if a[p, p] = 0 then
                    begin prtext ('failure') ; goto ex end ;
alpha:         a[p, p] := 1/a[p, p] ;
beta:          for j := 1 step 1 until p−1, p+1 step 1 until n do
                    a[p, j] := − a[p, j] × a[p, p] ;
```

delta: **for** $i := 1$ **step** 1 **until** $p-1, p+1$ **step** 1 **until** n **do**
                    for $j := 1$ **step** 1 **until** $p-1, p+1$ **step** 1 **until** n **do**
                        $a[i, j] := a[i, j] + a[i, p] \times a[p, j]$;
gamma: **for** $i := 1$ **step** 1 **until** $p-1, p+1$ **step** 1 **until** n **do**
                        $a[i, p] := a[i, p] \times a[p, p]$
            end *loop* ;
    comment Follows output of inverted matrix column by column ;
out: **for** $j := 1$ **step** 1 **until** n **do**
            begin
                line ;
                for $i := 1$ **step** 1 **until** n **do**
                    begin *line* ; *print* $(a[i, j])$ **end** ;
            end j ;
ex: **end**
    end *of program* ».

43.3. Block structure and storage economy

The possibility of declaring quantities not only at the very beginning of a program but also at higher block levels, can be used to keep storage requirements at a minimum. In order to achieve a saving, array declarations are delegated as far as possible to higher block levels (which may have been constructed extra for this purpose). True, an appreciable saving results only if we succeed in finding at least two disjoint blocks in each of which at least one fairly big array is declared, e.g.

            «**begin**
                integer n ;
                array $a[1:50, 1:50]$;
                    ⋮
        x: **begin**
                    integer array $b[1:n, 1:2 \times n]$;
                        ⋮
                end x ;
        y: **begin**
                    array $c[-n:n, 1:n]$;
                        ⋮
                end y ;
            end».

Here, 2501 positions are used throughout the program[1] for n and the array a, while $2n^2$ further positions are used only in block x, and

---

[1] Without counting storage used by the object program for organisational purposes.

$2n^2+n$ positions only in block y. The total storage requirement for all visible quantities is therefore $2n^2+n+2501$. However, for the program

```
«begin
    integer n ;
    array a[1:50, 1:50] ;
        ⋮
x:  begin
        integer array b[1:n, 1:2×n] ;
            ⋮
z:      begin
            array b[−n:n, 1:n] ;
                ⋮
        end z
    end x
end»,
```

the storage reserved for array b which is suppressed in block z is not set free after entry to this block, and therefore the total storage requirement is $4n^2+n+2501$ in this case.

Chapter VII

Procedures

A procedure is an operator which has the operational characteristics of a block and therefore also the same classes of operands (arguments, results, transients, exits) as a block. Unlike a block, however, some of the operands of a procedure may be exchangeable.

Procedures are either fixed constituents of the language (standard functions and standard I/O-procedures) or defined by procedure declarations.

§ 44. Procedure Declarations II

The following text describes the semantics of *ordinary procedures*, but with the amendments to be given later in § 46, it is also valid for *function procedures*.

44.1. Introduction

Given an arbitrary statement S which performs a certain action, we can easily turn it into the declaration of a procedure which performs exactly the same action. Take for instance the statement

«**begin**
   integer k ;
   $s := 0$;
   for $k := 1$ **step** 1 **until** n **do** $s := s + x[k] \times y[k]$
end»,

which computes the inner product s of given vectors $x[1:n]$, $y[1:n]$.

44.1.1. By placing an appropriate *procedure heading* in front of it, e.g.

«**procedure** *inner* ;
begin
   integer k ;
   $s := 0$;
   for $k := 1$ **step** 1 **until** n **do** $s := s + x[k] \times y[k]$
end»,

we obtain immediately the declaration of a procedure *inner* having the property that a corresponding procedure statement «*inner*» causes execution of the originally given statement and thus computes the inner product s of the two vectors $x[1:n]$, $y[1:n]$. In this procedure the four names n, x, y, s always denote the same quantities (but their values may of course change) which therefore are *predetermined operands* of the procedure.

44.1.2. A programmer may find little use for a procedure which cannot compute inner products of vectors with names other than x and y. In order to remove this restriction, the above procedure declaration can be modified into

«**procedure** *inner* (x, y) ; **array** x, y ;
  begin ⎫
       ⎬ procedure body as in 44.1.1.
  end». ⎭

Here x, y are *formal parameters*, i.e. identifiers representing only formal operands of the procedure (the true operands being designated only upon call of the procedure). In other words, the procedure can now be used to compute inner products of arbitrarily named vectors; for instance, the procedure call «*inner* (a, b)» will compute the inner product s of the vectors $a[1:n]$, $b[1:n]$ in precisely the same way as the procedure body operates on x and y.

On the other hand, n and s are still predetermined operands; to compute the inner product x of the vectors $p[1:k]$, $q[1:k]$, for instance, the sequence

$$«n := k ;$$
$$inner (p, q) ;$$
$$x := s»$$

would then be required.

44.1.3. A still more flexible procedure is obtained if also n and s are listed as formal parameters[1]:

«**procedure** *inner* (n, x, y, s) ; **value** n ;
    real s ; **integer** n ; **array** x, y ;
    begin ⎫
         ⎬ procedure body as in 44.1.1.
    end». ⎭

After these modifications we can use procedure *inner* to compute inner products of arbitrary vectors with arbitrary upper subscript bounds and can assign the resulting value to an arbitrary variable. As an example

«*inner* (k, p, q, x)»   computes the inner product x of the vectors $p[1:k]$, $q[1:k]$, while

«*inner* (10, z, z, *zeta*)»   computes the inner product *zeta* of $z[1:10]$ with itself, etc.

---

[1] For the meaning of «**value** n ;» see 44.6, 45.3.4.

44.2. Operands of a procedure

The operands of a procedure, i.e. the quantities involved in its execution, are essentially the operands of the *fictitious block* which — if S stands for the procedure body — is defined as the construction

«begin real æ ; S end»[1,2].

Indeed, execution of a procedure means essentially execution of this fictitious block, as we shall see later. However, it is one of the most important properties of procedures that their operands — besides being distinguished as arguments, results, transients and exits — fall into three categories, namely[3]

a) Those operands of the fictitious block whose identifiers are *not* quoted in the formal parameter part are called *global operands* of the procedure. The latter are operands of a procedure in the same sense as quantities are operands of a block.

b) Those operands of the fictitious block whose identifiers *are* listed in the formal parameter part[4] are called *formal operands* of the procedure. They are exchangeable in the sense that upon a procedure call other quantities are designated to be used in place of them.

c) In addition a procedure may have *hidden operands* like a block.

44.3. Rules for global parameters

44.3.1. A global parameter — that is, the identifier of a global operand — represents the same quantity inside the procedure body as outside in the environment of the procedure declaration. A global operand is therefore simply the extension of a quantity which exists outside the procedure. As a consequence we have

44.3.2. *The environment rule for global parameters*:

If the identifier I is global parameter of a procedure, then a (true or formal) quantity Q with that identifier must exist in the environment of the procedure declaration, and it is this Q which in a call of the procedure is meant by the identifier I.

---

[1] The declaration of the fictitious variable æ serves solely to make this piece of program a block.

[2] In case S is already an unlabelled block, this artificial construction is unneeded and we could take S instead. However, the fictitious block is necessary to cover the most general case, namely in order to avoid a destination label inside or in front of S from being mistaken for an operand of the procedure (which it may never be according to 5.4.3 of the RAR).

[3] We have chosen the terms global and formal *operands* to denote the quantities, whereas global and formal *parameters* are reserved to denote their identifiers.

[4] Usually only quantities actually occurring in the procedure body are quoted as formal operands, but this is not a strict rule. Indeed, circumstances may force a programmer to introduce formal operands which are not actually used in the procedure body (dummy operands). Examples of this kind occur in § 48.

According to this rule, a global parameter acts like a thread which links the procedure declaration permanently to its environment; indeed, a procedure which has global parameters is only fully defined if it is embedded into an ALGOL program in which the global operands are properly declared.

44.3.3. Consider, for instance, procedure *inner* as declared above in 44.1.1. It has the global operands n, x, y, s because these quantities appear inside the procedure body, neither being local to it nor being quoted as formal operands. As a consequence, this declaration can appear only at a place where quantities n, x, y, s of appropriate kinds (cf. 9.1) and types exist, e.g.:

« **begin**
    real s ; **integer** n ;
    $n := entier\,(z \uparrow 2 + 1)$;
l1: **begin**
        array $x, y\,[1:n]$;
        procedure *inner* ;
        begin ⎫
                ⎬ procedure body as in 44.1.1.
        end ; ⎭
          ⋮
    end *l1*
end».

Indeed, the environment of procedure *inner* is the floor of block *l1*, and in the floor of the latter the required quantities n, s, x, y correctly exist.

44.3.4. On the other hand, the following is an incorrect embedding:

«*l1*: **begin**
        real s, p, q, x, y ;
        integer n ;
        procedure *inner* ;
        begin ⎫
                ⎬ procedure body as in 44.1.1.
        end *inner* ; ⎭
        $n := z + 1$;
l2:   **begin**
        array $x, y\,[1:n]$;
          ⋮
        end *l2* ;
    end *l1*».

The environment of procedure *inner* is the floor of block *l1*, but since in this floor no arrays *x*, *y* can exist as would be required by the environment rule, the above piece of program is incorrect. The situation could be corrected, however, by moving the declaration for procedure *inner* to the floor of block *l2*.

44.3.5. *On the ordering of declarations.* If a procedure and some of its global operands (if any) are declared in the same block head, then the rule 42.3.2 requires that the declarations for the global operands are placed *before* the procedure declaration. Thus, if the incorrect example given in 44.3.4 is corrected by moving the declaration for procedure *inner* into block *l2*, it must be inserted there *after* the declaration for the two arrays *x*, *y*.

A further example:

```
«begin
    integer n ;
    procedure nix ; n := 0 ;
    procedure alto (ph) ; label ph ;
        begin nix ; goto ph end ; ...».
```

Since obviously *nix* is a global operand of procedure *alto*, and *n* is a global operand of *nix*, the declarations for these three quantities cannot be ordered other than shown in this example.

44.4. Rules for formal parameters

44.4.1. A formal parameter of a procedure X does not represent a true quantity as identifiers usually do but is only a designation given to a formal quantity (formal variable, formal array, formal string, formal label, formal switch, formal procedure) which exists only inside the procedure declaration without actually being declared there.

Moreover, a formal parameter has no connection whatsoever with any quantity having the same identifier and existing outside the procedure declaration, and it does not induce any requirements concerning the existence of certain quantities outside the procedure declaration as global parameters do.

The attributive *formal* has the meaning that such a quantity Q (except if called by value, for which see 44.6) has no independent existence as a declared quantity has but is only the representative for another quantity A which ultimately will be used as *actual operand* in place of Q.

44.4.2. *Specifications.* Whereas kind and type of a global operand are defined through the existence of the same quantity outside the procedure declaration, kinds and types of formal operands must be defined in the procedure declaration. To this end we have the *specification part*

$$«Z F, F, \ldots, F \; ; \; Z F, F, \ldots, F \; ; \; \ldots \; ; \; Z F, F, \ldots, F \; ; »^1,$$

in which the F's are the formal parameters of the procedure and the Z's are *specifiers* (for a complete list of possible specifiers see 41.2.1).

In the SUBSET the following rule must be observed (see SR, item 5.4.5):

> Every formal parameter F of a procedure must appear exactly once in the specification part.

A *single specification* $«Z \, F_1, F_2, \ldots, F_c \; ;»$ resembles a declaration somewhat; it declares the formal operands $F_1, \ldots, F_c$ to be of kind and type $«Z»$. Unlike declarations, however, specifications contain only the naked identifiers and therefore do not define additional properties such as subscript bounds of a formal array or the complete definition of a formal procedure (in fact, the absence of such additional information about formal operands adds to the flexibility of the procedure concept). On the other hand also formal labels and formal strings must be specified.

For procedure *inner* as declared in 44.1.3 the specification part is

$$«\textbf{real } s \; ; \; \textbf{integer } n \; ; \; \textbf{array } x, y»,$$

which says that x, y represent real arrays while n and s are simple variables of integer and real type respectively. This means that the actual operands to be used later in place of x, y, s must have these same respective kinds and types (this is not required for n since n is called by value, for which see 44.6).

44.4.3. *Structurized procedure headings*. As already indicated in § 41 and shown by some of the examples, the formal and actual parameter parts may be structurized. For procedure declarations to be published in the later volumes of this Handbook or distributed by the ALCOR group, it is recommended to structurize the formal and actual parameter part of procedure declarations and the corresponding calls in such a way that the four categories of operands (arguments, transients, results, and exits) are exhibited for the benefit of the reader.

The formal parameter part will therefore have the following syntactic form:

$$«(F, F, \ldots, F) \underbrace{trans \colon (F, F, \ldots, F)}_{\text{may be missing}} \underbrace{res \colon (F, F, \ldots, F)}_{\text{may be missing}} \underbrace{exit \colon (F, F, \ldots, F)}_{\text{may be missing}} ».$$

---

[1] For procedures which have no formal parameters the specification part is empty.

As an example, the formal parameter part in

«**procedure** *mica* (*a, b, c*) *trans*: (*d, e*) *res*: (*f*) *exit*: (*g, h*) ;
    real *c* ; **integer** *e* ; **array** *f* ; **Boolean array** *d* ;
    label *g* ; **switch** *h* ; **procedure** *b* ; **string** *a* ; ...»

indicates that procedure *mica* has the following operands:

1) String *a*, procedure *b*, and the real variable *c* are arguments.
2) Boolean array *d* and the integer variable *e* are transients.
3) The real array *f* is result.
4) Label *g* and switch *h* are exits.

For corresponding calls the same structurization of the actual parameter part is recommended (but not strictly required), e.g.

«*mica* ('012⊔*p*', *equ, v*) *trans*: (*d, ec*) *res*: (*ph*) *exit*: (*arica, pt*)».

Note that since all formal operands of a *function procedure* must be arguments, the structurization reduces for these to the first group.

44.4.4. *Predetermined versus exchangeable operands.* From what has been said above, it would seem that it is always an advantage to quote the operands of a procedure in the formal parameter part. Indeed, a procedure which has global operands is always somewhat hampered in its applicability since the global operands have predetermined names and must comply with the environment rule, while procedure declarations with only formal operands (so called *independent procedures,* for which see 47.1) are not linked to a specific environment but may be inserted into any block head of a program.

On the other hand, it must be pointed out that it serves no purpose to quote operands of a procedure as formal unless it is intended to make use of their exchangeability or to avoid the need for observing the environment rule for global parameters. To the contrary, the higher flexibility of formal operands must be paid for with a somewhat longer computing time as compared to global operands. It is therefore recommended to make operands global where they are predetermined by their nature and no conflict with the environment rule must be feared.

As an example, if at many places in a program the variables *a, b, c, d, g*[1], *g*[2], ..., *g*[*n*], but always only these, must be set to zero it is certainly most appropriate to declare

```
«procedure zero ;
begin
   integer k ;
   a := b := c := d := 0 ;
   for k := 1 step 1 until n do g[k] := 0
end zero».
```

Declaring this procedure as

«**procedure** *zero* (*n*) *res* : (*a*, *b*, *c*, *d*, *g*) ;
    real *a*, *b*, *c*, *d* ; **integer** *n* ; **array** *g* ;
    begin
            } procedure body as above
    end»

would have the only advantage that no environment rule need be observed, but it would be less efficient in use.

44.5. Scopes and procedure declarations

44.5.1. In § 42 we have excepted the interior of procedure declarations from all considerations of scopes. However, there are also rules for scopes applying inside procedure declarations. To discuss these, we again resort to the fictitious block which we imagine for a moment as being inserted in place of the procedure declaration:

a) Formal operands are considered as being local to the fictitious block as if they were declared there.

b) Global operands are true quantities existing outside the procedure declaration, whose scopes extend into the procedure body.

c) All other quantities occurring in the procedure declaration are either local to the fictitious block or to one of its subblocks; these are called the *internal quantities* of the procedure.

d) In the fictitious block the concepts *local, global, suppressed* apply as usual for true as well as for formal quantities, but procedure declarations contained in the procedure body are again considered separately.

44.5.2. According to these rules, a formal quantity may be suppressed inside the procedure body (though again this is not recommended); the *suppressor* is then no longer an operand but a *true quantity* local to some subblock of the procedure body. Accordingly, the identifier of the suppressing quantity is no longer considered as a formal parameter of the procedure. Furthermore, a formal operand *F* may occur as global operand of a procedure *Y* which is declared in the body of procedure *X*; in this case the scope of *F* extends into the body of procedure *Y*. Thus on the whole the same peculiarities may occur with formal quantities as with true quantities.

A peculiar situation arises if a programmer erroneously quotes a local quantity of the procedure body as formal operand, e.g.

«**procedure** *xxx* (*a*, *b*) ;
    real *a*, *b* ;

begin
   integer a ;
   $\vdots$
end xxx ».

This is not illegal, but since the formal operand a is suppressed through-out the procedure body, it cannot be involved at all in the execution of the procedure and is therefore a *dummy operand* of xxx.

44.5.3. However, it is illegal if a formal parameter coincides with a destination label that is located *outside* the biggest block occurring in the procedure declaration (e.g. a destination label in front of the procedure body):

  «**procedure** $yyy\,(a, b)$;
    real a, b ;
a: **begin** ... **end** yyy».

Indeed, the general rule given in 42.3.1 is violated in that case since both the label a and the formal operand a are local to the fictitious block.

44.6. The value part

44.6.1. The designer of a procedure has the possibility to list certain formal parameters in the value part

$$\text{«\textbf{value} } F, F, \ldots, F \text{ ;»}^{[1]}$$

of the procedure heading. The formal parameters (operands) thus quoted in the value part are said to be *called by value*, while all other formal parameters (operands) are said to be *called by name*.

44.6.2. If a formal operand X is called by value, this has the consequence that it is considered as a *true quantity local to the fictitious block* as if it were declared there. Then upon a procedure call, not a quantity, but a *value* (single value or an array of values) is designated as corresponding operand, and this value is assigned to X just prior to the execution of the procedure.

44.6.3. The precise effect of calling a formal operand by value can be defined only in connection with a procedure call, for which see § 45, but let us mention some of the consequences drawn from the rules given there:

   a) Only formal operands specified as *simple variables* or *arrays* can be called by value since only such quantities can have values.

---

[1] The value part is empty if none of the formal operands is called by value, and a fortiori if the procedure has no formal parameters.

b) A formal operand X which is called by value is automatically an argument of the procedure since any changes that X might formally undergo in the procedure body cannot feed back to the outside.

c) Formal operands intended as *transients* or *results* of the procedure should therefore not be called by value. Indeed, if procedure *inner* as declared in 44.1.3 were modified into

«**procedure** *inner* (n, x, y, s) ;
    value n, s ;
    real s ; **integer** n ; **array** x, y ;
    begin
          } procedure body as in 44.1.1.
    end»,

it would never be possible to get hold of the computed inner product s after execution of the procedure.

d) Formal parameters which represent *simple variables* and are arguments of the procedure, *should always be quoted in the value part*. This does not modify the effect of the procedure, but it is the only way to allow an expression to be used as corresponding actual parameter (see SR, item 4.7.3.2).

44.7. Further examples of procedure declarations[1]

44.7.1.

«**procedure** *polar* ;
    begin
        $r := sqrt(x{\uparrow}2 + y{\uparrow}2)$;
        $phi :=$ **if** $x=0$ **then** $90 \times sign(y)$ **else** $57.295779513 \times arctan(y/x)$
                $+$ (**if** $x{\geq}0$ **then** 0 **else if** $y{\geq}0$ **then** 180 **else** -180)
    end *polar*».

Obviously *polar* computes the polar coordinates r and *phi* (the latter in hexagesimal degrees but with decimal fractions) of a point whose cartesian coordinates are given as x, y. The operands x, y, r, phi are all predetermined.

44.7.2. The following procedure *matvec* serves to multiply a $m \times n$-matrix $a[1:m, 1:n]$ with a vector $x[1:n]$ yielding a vector $y[1:m]$. Here, all operands m, n, a, x, y are formal:

«**procedure** *matvec* (m, n, a, x) *res*: (y) ;   { Procedure identifier, formal parameter part.
    value m, n ;   Value part.
    integer m, n ; **array** a, x, y ;   Specification part.

---

[1] Some of these examples are derived from programs given in Chapter V.

```
begin                              Beginning of procedure body.
   real s ;
   integer i, k ;
   for i := 1 step 1 until m do
   begin
      s := 0 ;
      for k := 1 step 1 until n do s := s + a[i, k] × x[k] ;
      y[i] := s
   end i ;
end matvec».
```

44.7.3. The following example is intended for use with the Adams-Bashforth method described in 33.2, where a call of a procedure *equ* occurs which serves to define the differential system to be integrated. With *equ* as declared below, 33.2 integrates the one-dimensional heat equation for a homogeneous bar with temperature kept zero at one but linearly rising at the other end. The differential equation system has been obtained by discretisation only in the space dimension (method of SLOBODJANSKI):

«**procedure** $equ(x, y, n)$ res:(z) ; $\left\{\begin{array}{l}\text{Procedure identifier, formal} \\ \text{parameter part.}\end{array}\right.$

    real x ; **integer** n ; **array** y, z ;   Specification part.

    comment *global operand*: t ; $\left\{\begin{array}{l}\text{A comment appended to the last} \\ \text{semicolon of the specification part} \\ \text{indicates that a quantity } t \text{ occurs as} \\ \text{global operand of this procedure.}\end{array}\right.$

```
   begin
      integer k ;
      z[1] := − 2 × y[1] + y[2] ;
      for k := 2 step 1 until n − 1 do
              z[k] := y[k − 1] − 2 × y[k] + y[k + 1] ;
      z[n] := y[n − 1] − 2 × y[n] + x × t
   end equ».
```

44.7.4. Neville interpolation (cf. 32.1):

«**procedure** $nevint(n, a, b, x)$ res:(f) ;
    value n, x ;
    real x, f ; **integer** n ; **array** a, b ;
    comment *nevint* computes for given values $a[i], b[i]$ $(i=0, 1, \ldots n)$ the value $f = f(x)$ of the unique interpolation polynomial of order n, which is defined by $f(a[i]) = b[i]$ $(i = 0, 1, \ldots n)$;
    begin
       integer j, k ;
       array $y[0:n]$;

$btoy$: **for** $j := 0$ **step** 1 **until** n **do** $y[j] := b[j]$;
  for $k := 1$ **step** 1 **until** n **do**
   for $j := n$ **step** -1 **until** k **do**
    $y[j] := y[j] + (x - a[j]) \times (y[j] - y[j-1])/(a[j] - a[j-k])$;
  $f := y[n]$
 end $nevint$ ».

In order to avoid destruction of the value of the formal array b, its value is first assigned to an array y which is local to the procedure body (statement $btoy$), and all necessary operations are performed with this y. This way of saving the array b is more economical with respect to storage space than calling it by value. Indeed, our method requires the absolute minimum of storage space for achieving the purpose, whereas the storage space required if b is called by value is *at least* that much (cf. 45.2.1).

44.7.5. Inversion of a square matrix $a[1:n, 1:n]$ "on the spot", i.e. by a method which uses the storage taken by the given array a as working storage for the inversion and for the inverted matrix. The method of inversion is the same as in 43.2.4, only that in case of a zero division a jump to the formal label *fail* occurs:

« **procedure** $matinv(n)$ $trans$: (a) $exit$: $(fail)$;
 value n ;
 integer n ; **array** a ; **label** $fail$;
 begin
   integer i, j, p ;
   for $p := 1$ **step** 1 **until** n **do**
$loop$: **begin**
    if $a[p, p] = 0$ **then goto** $fail$;
$alpha$: $a[p, p] := 1/a[p, p]$;
$beta$: **for** $j := 1$ **step** 1 **until** $p-1, p+1$ **step** 1 **until** n **do**
     $a[p, j] := -a[p, j] \times a[p, p]$;
$delta$: **for** $i := 1$ **step** 1 **until** $p-1, p+1$ **step** 1 **until** n **do**
    for $j := 1$ **step** 1 **until** $p-1, p+1$ **step** 1 **until** n **do**
     $a[i, j] := a[i, j] + a[i, p] \times a[p, j]$;
$gamma$: **for** $i := 1$ **step** 1 **until** $p-1, p+1$ **step** 1 **until** n **do**
     $a[i, p] := a[i, p] \times a[p, p]$
  end $loop$
 end $matinv$ ».

However, since this procedure simply takes the diagonal elements as pivots, its applicability is limited to such cases where the matrix $A = (a[i, j])$ is either symmetric and positive definite or diagonally dominant.

44.7.6. (This example makes use of the procedures *line, prtext* and *print* as defined in 43.2):

«**procedure** *remark* (x, l, s) ;
   value x, l ;
   real x ; **integer** l ; **string** s ;
   comment prints a remark given as a string s, then a value x, and
                 finally produces l blank lines ;
   begin
      integer k ;
      line ;
      prtext (s) ;
      print (x) ;
      for $k := 1$ **step** 1 **until** l **do** *line*
   end *remark*».

This procedure has the peculiarity that it does not use the formal string s directly but transfers it as an actual parameter to the call of procedure *prtext*.

44.7.7. (An example with a formal switch):

«**procedure** *logbra* (x) *exit*: $(excess, branch)$;
   value x ;
   real x ; **label** *excess* ; **switch** *branch* ;
   begin
      integer k ;
      if $x < 1 \lor x \geq 10$ **then goto** *excess* ;
      $k := 1 + entier\,(10 \times ln\,(x)/ln\,(10))$; $\left\{ \begin{array}{l}\text{Computation of interval} \\ \text{into which } x \text{ falls.}\end{array}\right.$

      if $k = 0$ **then** $k := 1$; $\left. \begin{array}{l}\text{Corrections made if round-} \\ \text{off errors produce a } k \text{ be-}\end{array}\right.$
      if $k = 11$ **then** $k := 10$;   yond the interval $1 \leq k \leq 10$.

      goto *branch* $[k]$
   end *logbra*».

This procedure serves to initiate — depending on the value of x — a jump to one of ten (as of yet unspecified) labels; these will be specified only at call time as entries in the declaration for the actual counterpart of the formal switch *branch* (cf. 45.6.3). The ten said labels correspond to the ten intervals $1 \leq x < 10^{0.1}$, $10^{0.1} \leq x < 10^{0.2}$, ..., $10^{0.9} \leq x < 10$; but if x is outside the interval $1 \leq x < 10$, a jump to *excess* occurs.

§ 45. Procedure Statements II

A procedure statement initiates execution of an ordinary procedure which previously has been declared. In addition the procedure statement defines — through its actual parameters — the operands to be used in this execution.

The present section defines the precise action of a procedure statement and insofar gives a dynamic description of the operational characteristics of a procedure (as an alternative to the more static definition of a procedure as given in § 44 above).

45.1. The actual-formal correspondence

By confronting the actual parameter part $(A_1, A_2, \ldots, A_p)$ of a procedure statement with the formal parameter part $(F_1, F_2, \ldots, F_p)$ of the corresponding procedure declaration, we obtain a one-to-one correspondence between the two sets which defines for every formal parameter F_k the *corresponding* actual parameter A_k and vice versa. We shall also say that A_k is the *actual counterpart* of F_k, and F_k the *formal counterpart* of A_k.

Let us exhibit the actual-formal correspondence for the call «*nevint* $(n + 1, arg, fct, 1.52, f)$» of procedure *nevint* as declared in 44.7.4:

$$
\begin{array}{lcccccc}
\text{formal parameter part:} & (n, & a, & b, & x) & res: (f) \\
& \updownarrow & \updownarrow & \updownarrow & \updownarrow & \updownarrow \\
\text{actual parameter part:} & (n + 1, & arg, & fct, & 1.52, & f)
\end{array}
$$

45.2. Execution of a procedure statement

The execution of a procedure statement is equivalent to the execution of a certain block as if it were inserted in place of the procedure statement. This block — sometimes called the *equivalence block* of the procedure statement — is essentially the fictitious block[1], but with all formal parameters called by name being replaced by the corresponding actual parameters. This entirely hypothetical construction may serve for determining the effect of a call of a given procedure but otherwise is never actually needed. The precise rules for constructing (syntactically) the equivalence block are as follows:

45.2.1. For every *formal parameter F called by value* a corresponding declaration is put into the floor of the fictitious block[2]. In addition, we insert after these declarations for every such F a statement which assigns to F the current value of the actual counterpart A.

As an example, for an F specified as **Boolean**, this requires only

$$\text{«\,\textbf{Boolean}\ } F \ ;\text{»} \quad \text{and} \quad \text{«}F := A \ ;\text{»}$$

---

[1] If the value part is empty and the procedure body is already an unlabelled block, the latter may be taken as fictitious block.

[2] If at least one such declaration is inserted into the floor of the fictitious block, the declaration for the dummy variable æ (cf. 44.2) can of course be omitted.

to be inserted at appropriate places in the floor of the fictitious block. However, if F is specified e.g. as **integer array**, the declaration becomes

«**integer array** $F[l1:u1, l2:u2, \ldots, ld:ud]$»,

where d and the l's and u's denote dimension and current array bounds of the *actual* array A, while the assignment must be described as

«**begin**
    integer $k1, k2, \ldots, kd$;
    for $k1 := l1$ **step** 1 **until** $u1$ **do**
    for $k2 := l2$ **step** 1 **until** $u2$ **do**
    $\vdots$
    for $kd := ld$ **step** 1 **until** ud **do**
        $F[k1, k2, \ldots, kd] := A[k1, k2, \ldots, kd]$
end».

In other words, the *whole* array A is copied onto the formal array F, which therefore is an array with the same array bounds as A. This latter fact is what may cause storage problems if an array is called by value, and therefore one should not do this light-heartedly.

45.2.2. *The substitution rule.* Wherever a formal parameter F *called by name* is found within the procedure body[1], it is replaced by the corresponding actual parameter (which in this case, according to the SR, item 4.7.3.2, must be an identifier or a string). It should be recognized that — with the possible exception of formal strings — this is a replacement of identifiers by identifiers regardless of their syntactic positions, that is, «$a[i, k]$» is replaced by «$bb1[i, j]$» if $bb1$ and j are the actual counterparts of a and k.

45.2.3. *Name conflicts.* Where an identifier already used as name of an internal quantity of the procedure coincides with an identifier to be inserted into the fictitious block by one of the manipulations described above in 45.2.1 or 45.2.2, the name of the internal quantity must be changed[2] before the said manipulations may take place.

Indeed, without this rule, the effect of a call «$x(z)$» of a

«**procedure** $x(y)$;
    real y ;
    begin real z ; $z := 2 \times y$; $y := y/z$ **end**»

---

[1] Note that where a formal operand is suppressed in a subblock of the procedure body, the identifier of the suppressing quantity is no longer formal parameter but just an ordinary identifier.

[2] We shall indicate such systematic name changes by appending the Danish letter æ to the identifier.

would erroneously be interpreted as

$$\text{«\textbf{begin real} } z \text{ ; } z := 2 \times y \text{ ; } z := z/z \textbf{ end}\text{»,}$$

which certainly was not the intention of the designer of the procedure. With the above amendment, however, the internal z is changed into $z\text{æ}$, after which we obtain the equivalence block correctly as

$$\text{«\textbf{begin real} } z\text{æ} \text{ ; } z\text{æ} := 2 \times z \text{ ; } z := z/z\text{æ} \textbf{ end}\text{».}$$

As another example, consider the call «$arcsin(c[d], e)$» of

> «**procedure** $arcsin(x, y)$; **value** x ;
>   **real** x, y ;
> d: $y := arctan(x/sqrt(1 - x\uparrow 2))$ »,

which with only the rules 45.2.1 and 45.2.2 would be interpreted as being equivalent to the block

> «**begin**
>   **real** x ;
> $x := c[d]$;
> d: $e := arctan(x/sqrt(1 - x\uparrow 2))$
>   **end**».

By virtue of the rule 42.3.1 this is obviously an illegal piece of program, but the above amendment corrects the situation by changing the label d into $d\text{æ}$.

45.2.4. *Suppressed global operands.* If global operands of a procedure are suppressed at the location of a call, then — in order to meet the requirements of the environment rule for global parameters — the identifiers of the suppressing quantities are changed systematically before the manipulations described in 45.2.1 and 45.2.2 take place[1]. Consider, for instance

> «**begin**
>   **integer** t ;
>   **procedure** $common(x)$; **real** x ; $t := x$;
>   ⋮
> z: **begin**
>   **real** t ;
>   ⋮
> $common(t)$;
>   **end** z
>   **end**».

---

[1] This is the intent of the second sentence in 4.7.3.2 of the RAR.

Here the integer type variable t is suppressed in block z, and therefore the actual parameter of the call «*common* (t)» refers to the real type variable which is local to block z. Notwithstanding, according to 44.3, the t occurring as global parameter of procedure *common* refers to the suppressed quantity t. The above rule makes this evident by requiring that the name of the real type variable t be changed throughout block z into $t\text{æ}$ before the substitution rule is applied:

> «**begin**
>   **integer** t ;
>   **procedure** *common* (x) ; **real** x ; $t := x$;
>   ⋮
> z: **begin**
>   **real** $t\text{æ}$;
>   ⋮
>   *common* $(t\text{æ})$;
>   **end** z
> **end**».

Now the substitution rule yields the equivalent block for the call «*common* (t)» correctly as (æ denoting again the hypothetical variable necessary to make this piece of program a block)

$$\text{«begin real } \text{æ} \text{ ; } t := t\text{æ end»}.$$

Accordingly, this call accomplishes something which would seem impossible, namely changing the value of a suppressed variable.

45.2.5. It should be recognized that 45.2.3 and 45.2.4 add nothing new to the language but simply express the fact that the manipulations described in 45.2.1 and 45.2.2 (which are only auxiliary processes for explaining the effect of procedure calls) can of course not change the identity of quantities, with the exception that the formal operands called by name are identified with their actual counterparts.

Consequently no name changes apply where the identifier of a global operand *not suppressed* at the location of a procedure call coincides with the identifier of an actual operand. Indeed, e.g. in

> «**begin**
>   **real** z ;
>   **procedure** $x(y)$; **real** y ;   ⎫ Declaration of
>   **begin** $z := 2 \times y$; $y := y/z$ **end** ;   ⎭ a procedure x.
>   ⋮
> $x(z)$ Call of x.
> **end**»

the z occurring as actual operand in «$x(z)$» and the z which appears as global operand of procedure x, are one and the same quantity. It is therefore in order that the equivalent block for the call «$x(z)$» becomes

«**begin real** $æ$; **begin** $z := 2 \times z$; $z := z/z$ **end end**».

(This is a special case of a Gauss-Seidel effect, for which see 45.5.)

45.3. Restrictions for actual parameters

The actual parameters of a procedure statement must meet certain conditions, most of which follow from the fact that insertion of the equivalence block in place of the procedure statement must produce a meaningful piece of program:

45.3.1. If a formal parameter F_k of a procedure X is specified as **string**, then its actual counterpart A_k must be either a string or an identifier I representing a formal string. In the second case I must be formal parameter of a procedure Y whose body contains the call of X. An example of this sort is the call «$prtext(s)$» occurring in the example 44.7.6, which serves there to print the text that on a call of *remark* is given as actual operand (string) in place of the formal s.

45.3.2. If F_k is otherwise *called by name*, then the corresponding actual parameter A_k must be an identifier representing a quantity of the same kind and type as F_k (cf. SR, item 4.7.3.2 and 4.7.5.5). It is therefore not allowed to use e.g. a real type variable as actual operand where the corresponding formal parameter is called by name and specified **integer**.

45.3.3. If F_k is *called by value but specified as an array* (**array, real array, integer array, Boolean array**), then also A_k must be an identifier representing an array, but the type of A_k needs only be compatible with the type of F_k, i.e.

F_k specified as	Type of A_k
real	**real** or **integer**[1]
integer	**real** or **integer**[1]
Boolean	**Boolean**

45.3.4. If F_k is *called by value and specified as a simple variable* (**real, integer, Boolean**), then the actual parameter A_k must be an expression, namely an arithmetic expression if F_k is specified **real** or **integer**, and a Boolean expression if F_k is specified **Boolean**. (Variable identifiers

---

[1] If needed, the rules of § 20 come into action in these cases.

appearing as actual counterparts of formal parameters called by value
are automatically interpreted as special cases of expressions.)

Note that the above case is the only one which allows subscripted
variables and numerical or logical constants as actual parameters; indeed, procedure *inner* as declared in 44.1.3 could not have been called
by a procedure statement «*inner* (10, z, z, zeta)» with a numerical constant as first actual parameter if the corresponding formal parameter n
had not been called by value.

45.3.5. As a special case of these rules we mention that procedure *nevint*
as declared in 44.7.4 is incorrectly called in the following piece of program:

> «**begin**
>> **integer array** $x[0:m]$;
>> $\vdots$
>> *nevint* (m, z, x, t) *res*: (f) ;
>> $\vdots$
> $\vdots$ ».

This violates rule 45.3.2 because x is an integer array, whereas the formal
counterpart b in the declaration of *nevint* is called by name and specified
as a real array. Our example would be correct, however, if the formal
parameter b had been called by value.

45.4. Additional rules for arrays, procedures, switches

45.4.1. Where formal operands of a procedure X have additional properties (e.g. dimension of a formal array, number of parameters of a
formal procedure, or number of entries in the switch list of a formal
switch), these properties must be shared by the corresponding actual
operands.

However, such properties of formal operands are never explicitly
stated in the procedure declaration but usually can be found by inspection of the procedure body. For instance, if a is formal operand of a
procedure X, and a subscripted variable «$a[i, j+k, 15]$» occurs in the
body of X, we know that a is a three-dimensional array. The array
bounds are much more difficult to determine, however, and in fact may
depend on other formal operands. Furthermore, a call «$z(p, q, r)$» of a
formal procedure z immediately shows that the latter has three operands.

45.4.2. Wherever such properties of formal operands can be determined,
the requirement that the equivalent block must be a meaningful piece
of program has the following consequences:

a) If F_k is a formal array (called by name or value), its actual counterpart A_k must be an array of the same dimension. Concerning the array
bounds, we only require that those of A_k are at least as wide as those of F_k.

b) If F_k is a formal procedure, the corresponding actual operand A_k must be a procedure with the same number of operands as F_k; moreover, the rules of this § 45 must hold between the actual operands of calls of F_k (such calls may occur in the body of X) and the formal operands in the declaration of A_k (which must be given somewhere outside X).

c) If F_k is a formal switch, then also its actual counterpart A_k must be a switch and its switch list must have at least as many entries as the switch designators corresponding to F_k require.

45.4.3. *Example.* Inspection of the body of procedure *euler* as declared in 48.1.3 reveals that the formal operands y and yy are one- and two-dimensional arrays with array bounds $[1:n]$ and $[1:p, 1:n]$ respectively. Furthermore, *fct* is a procedure having itself four formal operands of the respective kinds and types (from left to right): **real** (by value), **real array**, **integer** and **real array**. As a consequence, a call

$$\text{«} euler\,(0,\,2,\,_{10}{-}2,\,100,\,sel)\ trans\!:\!(ky)\ res\!:\!(ty)\text{»}$$

requires that the following objects have been declared:

a) «**real array** $ty\,[1\!:\!100,\,1\!:\!2]$, $ky\,[1\!:\!2]$» (these arrays may also have wider array bounds).

b) «**procedure** $sel\,(x, y, z, w)$; **value** x ;
      real x ; **integer** z ; **array** y, w ;
      begin ... **end**».

45.4.4. There are cases where inspection of the procedure body is either impossible (e.g. for code procedures, cf. § 47) or does not reveal further properties of the formal parameters. In such cases additional information about the formal operands may be needed and should then be given e.g. by a comment or through "directions for the use of the procedure". However, the total lack of clues as to the properties of formal operands may also mean that a), b), c) of 45.4.2 need not be observed at all for certain operands.

An example of the latter sort is the standard I/O-procedure *outarray* (cf. § 49), which allows to output arrays of arbitrary dimensions and array bounds. Another example is

    «**procedure** $vgs\,(phi, x, y)$;
      array x, y ; **procedure** phi ;
      comment uses global operands: p, q ;
      $phi\,(p, x, y, q)$».

Since the body of this procedure is a call of a formal procedure *phi*, the dimensions of the formal arrays x, y depend entirely upon the actual counterpart of *phi* to be used in a future call of *vgs*. For instance, if we call it by

$$«vgs\,(inner,\ a,\ b)»,$$

where the actual operand *inner* is the procedure which has been declared as an example in 44.1.3, then *a* and *b* must be one-dimensional arrays, but in other cases the actual counterparts of *x* and *y* may well be multi-dimensional arrays.

45.5. Gauss-Seidel effects

45.5.1. From what has been said so far, we would expect that a procedure statement performs upon the actual operands always exactly those operations which in the procedure declaration are performed upon the formal operands. However, this is unconditionally true only if the actual operands coincide neither with each other, nor with global operands, nor with hidden operands of the procedure.

It should be clear, however, that such coincidences are absolutely allowed as far as the language is concerned, but they may cause unexpected effects:

45.5.2. Consider for instance procedure *matvec* as declared in 44.7.2, which serves to compute the product *y* of a matrix *a* with a vector *x*. Assume now that the product of a high power of the matrix $a\,[1\!:\!n,\ 1\!:\!n]$ with a given vector $b\,[1\!:\!n]$ is to be computed. This can be achieved by generating a sequence of vectors

$$b^{(0)} = b,\ b^{(1)},\ b^{(2)},\ \ldots,\ b^{(m)},$$

where   (1)

$$b^{(k)} = a \times b^{(k-1)}.$$

Since only the last vector of this sequence is relevant, we may store all vectors $b^{(k)}$ in the same array $b\,[1\!:\!n]$, which then at any time contains the most recently computed vector $b^{(k)}$. With this convention, (1) can be rewritten symbolically as $b := a \times b$, which, as it would seem could be described in ALGOL most conveniently as

$$«matvec\,(n,\ n,\ a,\ b)\ res\!:(b)»\,.$$

This is a fully legal call of *matvec*, but according to the rules given in 45.2 the equivalent block for this call is

```
«begin
    integer næ, m ;
    næ := m := n ;
    begin
        real s ; integer i, k ;
        for i := 1 step 1 until m do
```

begin
    $s := 0$;
    for $k := 1$ **step** 1 **until** $n\mathit{\ae}$ **do** $s := s + a[i, k] \times b[k]$;
    $b[i] := s$
  end i
 end *matvec*
end *equivalence block*».

We recognize immediately that the components of the vector b to be multiplied by the matrix a are changed before the computation of $a \times b$ is completed. The net effect of this is that the vector b is not actually multiplied with the matrix $A = (a[i, k])$ but with

Fig. 38

Since this latter matrix is closely related with the Gauss-Seidel method[1] for solving linear equations, the undesired effects caused by coincidences among actual (and global) operands are called *Gauss-Seidel effects*.

To correct the above example, we may introduce an additional vector $b1[1:n]$ and then shuttle between b and $b1$:

    «*matvec* (n, n, a, b) *res*: $(b1)$;
    matvec $(n, n, a, b1)$ *res*: (b) ».

45.5.3. A less obvious example is

  «**begin**
    real a, b, c ;
    procedure *atobe* ; $b := a$;
    procedure *oneto* (x) ; **real** x ;
      begin
        $x := 1$;
        atobe
      end *oneto* ;
      ⋮
    oneto (b)
  end ».

---

[1] See § 23 in ZURMÜHL [*44*].

13 Rutishauser, Description of ALGOL 60

Here the actual operand b of the call «*oneto*(b)» is at the same time global operand of *atobe*, hence hidden operand of procedure *oneto*, which makes that call susceptible to a Gauss-Seidel effect. Indeed, the equivalence block for this call is

«**begin**
   $b := 1$;
   $b := a$
end».

Note, however, that this effect would not materialize if the hidden operand b of *oneto* were suppressed at the location of the call since then the two b's would in fact represent two different quantities.

45.5.4. *Safeguards against Gauss-Seidel effects.* In principle we could always check by inspection of the procedure body whether or not coinciding operands might cause Gauss-Seidel effects. However, this troublesome task may be avoided if the following facts are observed:

a) It is always harmless if an actual operand *called by value* coincides with another (actual, global or hidden) operand.

If, for instance, in procedure *matvec* (44.7.2) the formal operand x had been called by value, then the statement

«**for** $k := 1$ **step** 1 **until** m **do** *matvec* (n, n, a, b) *res*:(b) »

would indeed produce the desired result $b^{(m)}$.

b) Gauss-Seidel effects cannot occur if a group of coinciding actual (including global and hidden) operands are all arguments of the procedure.

As an example, in the call

«*inner* $(10, z, z, zeta)$ »

(cf. 44.1.3) the two formal parameters x, y which are identified with the same actual quantity z are, though called by name, both arguments of the procedure *inner*, as may be seen by inspection of the procedure body. Therefore no Gauss-Seidel effect can occur, and the above statement will produce the inner product *zeta* of the vector $z[1:10]$ with itself as expected.

45.6. Further examples of procedure statements and their interpretation

45.6.1.   «*nevint* $(15, n, k, 1.52)$ *res*:(x) ».

The corresponding equivalent block reads:

«begin

> **integer** $n\alpha$; **real** $x\alpha$;
> $n\alpha := 15$; $x\alpha := 1.52$;

{ Declarations in the head of the equivalent block for formal operands called by value (cf. 45.2.1). Because the identifiers x, n appear as actual parameters of the present call, they are changed into $x\alpha$, $n\alpha$.

> **begin**

{ Beginning of procedure body in which, because also k is actual parameter, the internal k is changed into $k\alpha$, whereupon a, b, f are replaced throughout by n, k, x.

> > **integer** $j, k\alpha$;
> > **array** $y[0:n\alpha]$;
> > **for** $j := 0$ **step** 1 **until** $y[j] := k[j]$;
> > **for** $k\alpha := 1$ **step** 1 **until** $n\alpha$ **do**
> > > **for** $j := n\alpha$ **step** -1 **until** $k\alpha$ **do**
> > > $y[j] := y[j] + (x\alpha - n[j]) \times$
> > > $\qquad\qquad (y[j] - y[j-1])/(n[j] - n[j-k\alpha])$;
> > $x := y[n\alpha]$
> **end** *nevint*
> **end** *equivalent block»*.

As might be foreseen, because *nevint* has no global operands and the given procedure statement meets the conditions stated in 45.5.4 above, it performs the desired action, namely Neville interpolation of order 15 with 16 points given as abscissae $n[i]$ and ordinates $k[i]$.

45.6.2. *«remark* $(k, 15, 'fix \sqcup it')$» (compare 44.7.6) is equivalent to

«begin

> **real** x ; **integer** l ;
> $x := k$; $l := 15$;
> **begin**
> > **integer** $k\alpha$;
> > *line* ;
> > *prtext* $('fix \sqcup it')$;
> > *print* (x) ;
> > **for** $k\alpha := 1$ **step** 1 **until** l **do** *line*
> **end** *remark*
> **end** *equivalent block»*.

Note that in this case the change of the name of the internal variable k into $k\alpha$ was not actually needed.

45.6.3. In order to use procedure *logbra* of 44.7.7, a switch with at least 10 entries must be declared as actual counterpart of the formal switch *branch*, e.g.:

«**begin**
    switch $l := $ *one, two, three, four, five, six, seven, eight, nine, ten* ;
  nine : *seven* : ... ;
       ⋮
  four : *six* : *eight* : ... ;
       ⋮
  ten : *logbra* $(t \uparrow 2)$ *exit*: *(five, l)* ;
  five : *one* : *two* : ... ;
   end *of block* ;
three : ...».

First we observe that switch l meets the requirements of the environ-ment rules for switches (40.4.2); then, in applying the rules of 45.2, we obtain the following block, to which the execution of the above piece of program is equivalent:

«**begin**
    switch $l := $ *one, two, three, four, five, six, seven, eight, nine, ten* ;
  nine : *seven* : ... ;
       ⋮
  four : *six* : *eight* : ... ;
  ten : **begin**
            real x ;
            $x := t \uparrow 2$;
            begin
                integer k ;
                if $x < 1 \lor x \geq 10$ **then goto** *five* ;
                $k := 1 + entier (10 \times ln (x) / ln (10))$;
                if $k = 0$ **then** $k := 1$;
                if $k = 11$ **then** $k := 10$;
                goto $l[k]$
            end *logbra*
        end *equivalence block* ;
  five : *one* : *two* : ... ;
   end *of block* ;
three : ...».

§ 46. Function Procedures and their Use

In this section the properties and restrictions of function procedures are defined through amendments to the rules for ordinary procedures and their calls as given in § 44 and § 45.

46.1. Function procedure declarations

46.1.1. A procedure can be declared as a function procedure if it produces *exactly one single value as result* while all other operands (including hidden operands) are strictly arguments of the procedure.

Declaring a procedure as a function procedure has the purpose that its only result can be used directly as a primary in an expression. The declaration must then have the syntactic form described in 41.2.2.

46.1.2. The rules for global and formal operands, specification part and value part apply as for ordinary procedures (cf. 44.3—6), but in addition the following rules must be observed:

a) The procedure identifier must occur within the procedure body as if it were a simple variable, but this fictitious variable — called the *principal result* of the procedure — may appear only on the left side of assignment statements, as may be indicated by the example

«**real procedure** *radius* (x, y) ; **value** x, y ; **real** x, y ;
$radius := sqrt (x{\uparrow}2 + y{\uparrow}2)$ ».

b) The principal result may neither be declared in the floor of the procedure body[1] nor be specified in the procedure heading; its type is defined solely by the declarator which appears in front of the procedure declaration.

c) A function procedure must be declared in such a way that corresponding function designators cannot have any other effect than producing a single value (for the precise content of this rule see 46.5.3).

46.1.3. Restriction c) above requires that, except for the principal result, all operands of a function procedure (including hidden operands in the sense of 42.5.2) be arguments; as a consequence, e.g. neither labels nor switches may occur as formal operands of a function procedure (and therefore also not as actual operands of function designators). The following is therefore illegal:

«**real procedure** *sentum* (a, b, c) ;
    value a ;
    real a, b ; **label** c ;
    if $a > 0$ **then** *sentum* $:= exp (b \times ln (a))$ **else goto** c ».

On top of this, rule c) likewise *disallows ordinary procedures as formal operands of a function procedure* (cf. 46.5.4), whereas they can be global operands, provided the hidden operands introduced in this way are again strictly arguments.

---

[1] The identifier of the principal result might of course be declared again in a subblock of the procedure body with the effect that the principal result is suppressed in this subblock according to the general rules for scopes.

46.1.4. An already declared ordinary procedure which has only one result and neither exits nor transients can be transcribed almost mechanically into a function procedure, as we show now for procedure *inner* as declared in 44.1.3:

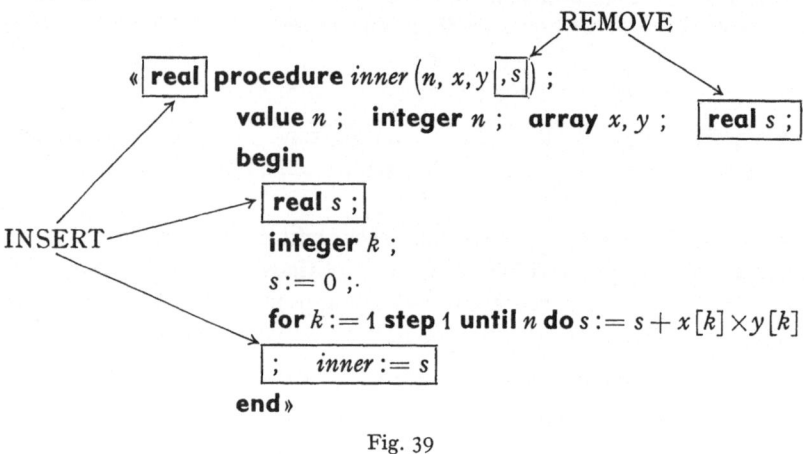

Fig. 39

46.2. Further examples of function procedure declarations

46.2.1. Example of a logical function:

«Boolean procedure *decide* (*n, v, w*) ;
    value *n, v* ;
    integer *n* ; **Boolean** *v* ; **Boolean array** *w* ;
    comment the principal result of this procedure obtains the value
              true if and only if all components $w[1], w[2], \ldots w[n]$
              are equivalent to v ;
    begin
        integer *k* ;
        Boolean *t* ;
        $t := $ **true** ;
        for $k := 1$ **step** 1 **until** n **do** $t := t \wedge (v \equiv w[k])$;
        decide $:= t$ { Assignment of function
    end *decide* ».   { value to the principal result.

The only operands (besides the principal result) are the formal quantities *n, v, w*, which we immediately recognize as being arguments. *Note* that in this example an auxiliary variable *t* had to be introduced for the sole reason that the principal result may never occur on the right side of an assignment statement.

46.2.2. «**real procedure** *zigzag*(*x*) ; **value** *x* ; **real** *x* ;

　　　　comment computes value of piecewise linear function as-
　　　　　suming values ± 1 at the integers $x = 4 \times n \pm 1$;

　　　　begin
　　　　　if $abs(x) > 2$ **then** $x := x - 4 \times entier(x/4 + 0.5)$;
　　　　　$zigzag :=$ **if** $abs(x) < 1$ **then** x **else** $2 \times sign(x) - x$
　　　　end *zigzag*».

Note that here *x*, because it is called by value, is an argument of the
procedure despite the fact that it appears within the procedure body on
the left side of an assignment statement.

46.2.3. (Compare also 31.4):

«**Boolean procedure** *stable* (*n, a*) ;
　　value *n, a* ;
　　integer *n* ; **real array** *a* ;
　　comment produces the value **true** if and only if all roots of the
　　　　polynomial $\sum_{0}^{n} a_k x^k$, given as an array $a[0:n]$, have nega-
　　　　tive real parts ;
　　begin
　　　　real *c* ;
　　　　integer *j, k* ;
　　　　$stable :=$ **false** ;
　　　　for $j := 0$ **step** 1 **until** $n - 1$ **do**
　　　　begin
　　　　　if $a[0] \times a[j+1] \leq 0$ **then goto** *ex* ;
　　　　　$c := -a[j]/a[j+1]$;
　　　　　for $k := j+2$ **step** 2 **until** $n - 1$ **do**
　　　　　　　$a[k] := a[k] + c \times a[k+1]$
　　　　end *j* ;
　　　　$stable :=$ **true** ;
ex: **end** *stable*».

Here the array *a* must be called by value in order to meet the requirement that it should be an argument of the procedure. Indeed, without
value call, *a* would be a transient since its components occur as assignment variables, and this would contradict 46.1.2, c. Furthermore, it
should be recognized that in this example two assignments to the principal result occur; in such a case the most recent of these assignments
defines the value of the function.

46.2.4. Example of an integer-valued function.

«**integer procedure** $gcd\,(m, n)$;
    value m, n ;
    integer m, n ;
    comment computes greatest common divisor of m and n ;
    begin
        integer c ;
        $m := abs\,(m)$;
        $n := abs\,(n)$;
        if $n=0$ **then goto** $zero$;
div: $c := entier\,(m/n)$;
        $m := m - c \times n$;
$rep1$: **if** $m \geqq n$ **then begin** $m := m - n$; **goto** $rep1$ **end** ; [1]
$rep2$: **if** $m < 0$ **then begin** $m := m + n$; **goto** $rep2$ **end** ; [1]
        $c := m$;
        $m := n$;
        $n := c$;
$zero$: **if** $n=0$ **then** $gcd := m$ **else goto** div
    end gcd».

46.3. Rules for function designators

While ordinary procedures are called through procedure statements, function procedures are called through function designators[2] which are constituents of arithmetic or Boolean expressions.

46.3.1. The *syntactic form of a function designator* is predetermined by the corresponding function procedure declaration, namely it is (compare 41.2):

«I»,   if procedure I was declared as
               «T **procedure** I ; S», and
«$I\,(A_1, A_2, \ldots, A_p)$», if procedure I was declared as
               «T **procedure** $I(F_1, F_2, \ldots, F_p)$; VCS».

In both these cases T stands for one of the declarators «**real**», «**integer**», «**Boolean**», while A and I have the same meaning as for procedure statements (cf. 26.2.1).

46.3.2. *Influence of scopes and types.* A function designator must be located within the scope of the corresponding function procedure and also within the scopes of all quantities appearing among its actual

---

[1] In the operation «$entier\,(m/n)$» the division m/n is subject to roundoff errors which may have the effect that the new m is outside the theoretical range $0 \leqq m < n$. The two statements labelled *rep1* and *rep2* serve to correct this occurrence.

[2] To be precise, also a function procedure might in principle be called through a procedure statement, but this has by definition no effect (see SR, item 5.4.4).

parameters, but otherwise it may occur wherever the rules for expressions allow

a) for a *primary* (cf. 16.2), in case the function designator is of real or integer type, or

b) for a *Boolean primary* (cf. 18.2), in case the function designator is of Boolean type.

46.3.3. *The actual-formal correspondence* between a function designator and a corresponding function procedure declaration is defined in exactly the same way as for ordinary procedures (cf. 45.1). Moreover, the restrictions for actual parameters of a procedure statement given in 45.3 apply analogously also for actual parameters of a function designator.

46.4. Evaluation of a function designator

46.4.1. Wherever a function designator is met during evaluation of an expression E, the actions which are taken are equivalent to the following operations:

a) The evaluation of E is interrupted and the equivalence block is constructed in the same way as described in 45.2 for procedure statements, but in addition the principal result is declared as local variable of the equivalence block.

b) This block is inserted in place of the function designator and executed.

c) When the «**end**» of the equivalence block is reached, the current value of the principal result is taken as the value of the function designator, and with that value the evaluation of E is continued while the equivalence block is discarded again.

d) If the execution of the equivalence block terminates through a jump to a destination lying outside, the further course of the computation is *undefined*.

46.4.2. Let us demonstrate what has been said above with the arithmetic expression

$$\text{« } p \times radius/sqrt\,(5+zigzag\,(p{\uparrow}2))\text{ »},$$

where the procedures *zigzag* and *radius* are those declared in 46.2.2 and 46.1.2 respectively. The steps to be taken are:

a) Execute the block

$$\text{«}\mathbf{begin\ real}\ radius\ ;\ radius := sqrt\,(x{\uparrow}2+y{\uparrow}2)\ \mathbf{end}\text{»}.$$

b) Execute the block

«**begin**
    real $x, zigzag$;
    $x := p{\uparrow}2$;

body:
```
   begin
      if abs (x) > 2 then x := x − 4 × entier (x/4+0.5) ;
      zigzag := if abs (x) < 1 then x else 2 × sign (x) − x
   end zigzag
end equivalence block».
```

c) Evaluate «$p \times radius/sqrt\,(5 + zigzag)$», where the values of the fictitious variables *radius* and *zigzag* are taken as they were produced by a) and b).

46.5. The side effect question

46.5.1. The original intention was that a function designator should evaluate a function which before had been defined by a corresponding declaration. Of course, it was agreed that in ALGOL the concept of a function should be somewhat more computer-oriented than in classical analysis, but it was felt that also in ALGOL a function should produce one single value to be used further in the evaluation of an arithmetic or Boolean expression.

However, in ALGOL 60 the declaration of a function procedure was given the same syntactic form as the declarations of ordinary procedures (except for a preceding type declarator), and this fact made constructions like the following possible in full ALGOL 60:

```
«real procedure se (a, b, c) ;
    real a ; integer b ; label c ;
    begin
       if a=0 then goto c ;
       b := entier (b/a) ;
       se := a × b
    end».
```

With this declaration, a function designator «$se\,(x, y, acryl)$» would, according to the rules of 46.4, initiate operations which are the equivalent to the following block

```
«begin
    real se ;
    begin
       if x=0 then goto acryl ;
       y := entier (y/x) ;
       se := x × y
    end se
 end equivalence block».
```

Accordingly, «*se*(*x*, *y*, *acryl*)» would not only produce a value *se* but would also change the value of *y* or cause a jump to *acryl*, and this during evaluation of an expression!

46.5.2. Such effects are called *side effects*, because the function designator does something *besides* producing a value. It should not be denied that with such side effects most elegant and useful effects might be achieved; however, for the less pretentious users of ALGOL this possibility is rather unwanted since side effects tend to disguise the intentions of the programmer, and ALGOL is a language for stating clearly rather than for disguising the intended computing process. In fact, the possibility of making jumps and assignments while evaluating an expression, hence also while evaluating subscripts and even array bounds of an array declaration, would open the door to confusion. For these reasons it was decided to disallow side effects in the SUBSET (cf. SR, item 5.4.4) by imposing the restriction 46.1.2,c. This restriction can be formulated more concisely as follows:

46.5.3. *The side-effect restriction.* A function procedure must be declared such that, if for any corresponding function designator meeting the requirements of 46.3 the equivalence block is constructed according to the rules given in 46.4, this block is equivalent to a dummy statement[1].

As an example, the equivalence block corresponding to the function designator «*se*(*x*, *y*, *acryl*)» given in 46.5.1 is obviously not equivalent to a dummy statement since it may produce the externally visible effect «*y* := *entier*(*y*/*x*)» or «**goto** *acryl*». Procedure *se* is therefore not allowed in SUBSET ALGOL 60.

On the other hand, the equivalence block associated with the function designator «*zigzag*(*p*↑2)» as shown in 46.4.2,b is indeed equivalent to a dummy statement because all it does is to make assignments to the variables *x* and *zigzag*, and these are local to the equivalence block.

46.5.4. *Ordinary procedures as formal operands of function procedures.* Consider, for instance,

  «**real procedure** *crit*(*n*, *a*, *xx*) ;
   value *n* ;
   integer *n* ; **array** *a* ; **procedure** *xx* ;
   begin
    real *d* ;
    xx(*n*, *a*, *d*) ;
    crit := *d*
   end *crit*».

---

[1] There would seem to be a contradiction between this rule and 46.4.1,c; however, here the equivalence block is executed without preserving the value of the principal result.

Here the intention is to compute a certain function of the components of the array $a[1:n]$, whereby an arbitrary procedure xx is involved. For a call «$crit(15, b, c)$» the equivalence block becomes

```
«begin
      real crit ;
      integer n ;
      n := 15 ;
      begin
          real d ;
          c(n, b, d) ;
          crit := d
      end crit
   end equivalence block».
```

Whether or not this block is equivalent to a dummy statement as the rule 46.5.3 requires, depends entirely upon the procedure c given as actual counterpart of xx in the call of $crit$; indeed, the global operands of c are hidden operands of the above block.

Now this block should be equivalent to a dummy statement for *any* choice of the procedure c, while on the other hand the programmer has the freedom to declare e.g.

«**procedure** $c(u, v, w)$; **integer** u ; **array** v ; **real** w ; **goto** zzz»,

where zzz occurs as destination label in the environment of the above declaration. Since in this way zzz has become hidden operand of $crit$, we recognize at once that the function procedure $crit$ violates the side-effect restriction, and so does any function procedure with an ordinary procedure as formal operand. As a *consequence function procedures cannot have ordinary procedures as formal operands.*

46.5.5. However, under certain conditions ordinary procedures are allowed as *global* operands of function procedures. As an example, the following piece of program is entirely legal:

```
«begin
    real x ;
    procedure xx(n, a) res:(d) ;
       value n ;
       real d ; integer n ; array a ;
       comment xx uses global operand x ;
       begin
          integer k ;
          d := 0 ;
          for k := n step −1 until 0 do d := d × x + a[k]
       end xx ;
```

real procedure *crit* (*n*, *a*) ;
   value *n* ;
   integer *n* ; **array** *a* ;
   comment *crit* uses global operand *xx* ;
   begin ⎫
             ⎬ procedure body as in 46.5.4
   end *xx* ; ⎭
  ⋮
   ».

Here *crit* has *xx* as global operand, but since the latter is predetermined and the only global operand of *xx* is *x*, which is an argument, no call of *crit* can violate the side-effect restriction.

§ 47. Code Procedures

The ALGOL report allows the body of a procedure to be written in a language different from ALGOL, e.g. in internal machine code or in an assembly language. This section deals with some of the practical consequences of this possibility.

47.1. Independent procedures

Procedures which have no global operands are called *independent*. Among the examples given so far, the procedures *matvec, nevint, matinv, remark, decide, zigzag, stable* and *gcd* are independent while *equ, polar* and *radius* are not.

An independent procedure is not linked to its environment; its declaration can be removed from its original location and inserted at another place in the program without changing its meaning. Even more, since the environment has no bearing on the effect of an independent procedure, it has become customary to write or to publish the corresponding procedure declaration without a surrounding ALGOL program. In fact, the present Handbook contains a collection of computing processes described as *independent procedure declarations* (this is the term for procedure declarations written out of context).

Obviously an independent procedure declaration is not actually a legal ALGOL program but only a construction ready for being copied into an ALGOL program. Such copying will immediately produce a legal ALGOL program provided the procedure identifier does not conflict with other names in the program.

47.2. Pseudodeclarations

According to the rules, an ALGOL program must contain the declarations of all procedures called within the program, except those for the standard functions and for the standard I/O-procedures. Thus, if one wishes to run a program using an independently declared procedure, the corresponding declaration must first be copied into that program (into the head of some block).

However, if the ALGOL program is to be published, such copying is most impractical; in fact it would mean that an already published procedure declaration is being published again. In order to avoid this waste, several philosophies have been introduced.

The ALCOR group decided that the declarations for independently declared procedures should be inserted at the proper places in the ALGOL programs in which they are used, but that hereby the procedure bodies might be abbreviated by the symbol «**code**», e.g.

> «**procedure** *matinv* (*n*) *trans* : (*a*) *exit* : (*fail*) ;
>   **value** *n* ;
>   **integer** *n* ; **array** *a* ; **label** *fail* ;
>   **code**»,
> «**real procedure** *zigzag* (*x*) ; **value** *x* ; **real** *x* ; **code**».

These "mutilated" procedure declarations used by the ALCOR group as representatives of independently declared procedures are called *pseudo-declarations*. Their syntactic form complies with the rules given in § 41 and § 44, only that the letter *S* now stands for the symbol «**code**», which, if we consider it as a piece of code (cf. RAR, section 5.4.6), is a fully legal construction.

The symbol «**code**» also has a practical significance: It informs the computer at compilation time that the actual procedure body can be found at some other place, e.g. in pretranslated form on a library tape.

Pseudodeclarations can also appear within procedure declarations, namely if they use previously declared procedures as subprocedures. For instance, if procedure *matvec* is already declared as shown in 44.7.2, a procedure *raylqu* for computing the Rayleigh quotient of a vector $x[1:n]$ with respect to a matrix $a[1:n, 1:n]$ can be simplified as follows:

> «**real procedure** *raylqu* (*n*, *x*, *a*) ;
>   **value** *n* ;
>   **integer** *n* ; **array** *x*, *a* ;
>   **begin**
>   **integer** *k* ;
>   **real** *g*, *h* ;
>   **array** *y* [1 : *n*] ;
>   **procedure** *matvec* (*m*, *n*, *a*, *x*) *res* : (*y*) ;
>   **value** *m*, *n* ;
>   **integer** *m*, *n* ; **array** *a*, *x*, *y* ;
>   **code** ;

program:
```
      g := h := 0 ;
      matvec (n, n, a, x) res: (y) ;
      for k := 1 step 1 until n do
      begin
          g := g + x[k]↑2 ;
          h := h + x[k] × y[k]
      end k ;
      raylqu := h/g
  end raylqu».
```

Of course not much ink is saved in this little example, but the use of previously declared procedures may allow considerable simplifications in other cases. Furthermore, it should be recognized that the object program corresponding to procedure *raylqu* requires indeed less space on the library tape in this way, and this is not offset by the space used for procedure *matvec*, since the latter is very likely used also upon other occasions.

Of course, the pseudodeclaration for procedure *matvec* could be removed from the body of *raylqu*, but then *matvec* would be a global operand of *raylqu*, and hence the latter would not be independent.

47.3. Code procedures

A code procedure is a procedure (function or ordinary) whose body is not written in ALGOL but in some other language, e.g. in internal machine code. Code procedures are needed for performing strongly computer-oriented operations such as in- and output or operations upon single bits of machine words, in short: for operations that cannot (or only inefficiently) be described in terms of ALGOL.

The heading of a code procedure declaration must obey the very same rules as for non-code-procedures, but the procedure body is given as a piece of code for which no rules can be given. In fact, the design of this piece of code is dictated entirely by the properties of the computer/compiler configuration with which the program is to be run.

As an example, we exhibit a procedure which performs for given integer values x, y, z (all $< 2^{46}$ in modulus) the decomposition of

$$x \times y + z \quad \text{into} \quad c \times 2^{46} + d,$$

where c, d are also $< 2^{46}$ in modulus. The body of this procedure is written in CODAP[1] [11], which is the assembly language for the CDC

---

[1] Some operation symbols of CODAP are: SLJ: jump; LDA: Clear and add; STA: Store; RTJ: Jump with automatic return; BSS: Reservation of a storage area; SAU: Substitute address of upper half word; SAL: Substitute address of lower half word. For further details see [11].

1604 A computer:

«**procedure** *precmp* (*x*, *y*, *z*) *res*: (*c*, *d*) ;
    integer *x*, *y*, *z*, *c*, *d* ;

PRECMP	SLJ		**	⎰ Entry point. After entry, return address
	LDA	6	0	⎱ is inserted in place of **.
	STA		X	
	LDA	6	1	
	STA		Y	The CDC 1604 A ALGOL compiler places
	LDA	6	2	links between actual and formal operands
	STA		Z	into consecutive positions *p*, *p*+1, *p*+2,
	LDA	6	3	..., where *p* is the contents of index reg-
	STA		C	ister 6. These links are now being stored
	LDA	6	4	into symbolic positions X, Y, Z, C, D.
	STA		D	
	RTJ		X	
X	BSS		1	
	SAU		MULT	
	RTJ		Y	
Y	BSS		1	
	SAL		MULT	
	RTJ		Z	Jumps to the program which calls *precmp*
Z	BSS		1	(with automatic returns) for fetching ad-
	SAL		MULT+3	dresses of actual operands; these addres-
	SAU		ADD−1	ses are inserted at appropriate places
	RTJ		C	below (indicated by **).
C	BSS		1	
	SAU		ADD+4	
	RTJ		D	
D	BSS		1	
	SAL		ADD+2	
MULT	LDA		**	
	MUF		**	
	LLS		1	
	STA		C	
	ALS		2	
	LRS		2	
	STQ		D	
	LDA		**	
	LDQ		C	
	QJP	Z	ADD	
+	SSK		D	
	SLJ		POS	

NEG	AJP	M ADD
	RAO	C
	LAC	BM
	SLJ	ADD−1
POS	AJP	P ADD
	RSO	C
	LDA	BM
+	ADD	**
ADD	ADD	D
	LRS	46
	STA	D
	ARS	1
	LRS	2
	STQ	**
	LDA	C
	ADD	D
	STA	**

These instructions describe the calculation proper of c, d from the given x, y, z.

** mean addresses to be inserted by the previous parts of the program, while C, D are auxiliary symbolic positions for intermediate results.

	SLJ	PRECMP
BM	OCT	2000000000000000 »

Jump back to first instruction, where exit from *precmp* occurs.

($2\uparrow46$ in octal form).

Obviously such a code procedure body has significance only for a specific computer/compiler configuration and should therefore not appear in an ALGOL program since we expect the latter to be a computer-independent description of a computing process. Consequently the ALCOR group decided that in such cases the symbol «**code**» be written in place of the procedure body. In other words, code procedures are treated like independent procedures; in both cases only the corresponding pseudo-declaration will appear in the ALGOL program that uses the procedure, while the full procedure body may be e.g. on a library tape, written in machine code or in some symbolic language.

It remains to say what will happen with code procedures when ALGOL programs are exchanged between different computers. For procedures whose body can be written in ALGOL we can assume that this ALGOL text is also exchanged and may either be used directly or (if it would produce a too inefficient object program) at least serve as a model for designing the procedure body directly in machine code. However, in many cases we can but describe the effect of a procedure in plain words. For instance, in the case of a procedure declared as

«**procedure** *output* (a, b, c) ;
    value a, c ;
    real c ; **integer** a ; **string** b ;
    code»,

the description of the operational behaviour might read as follows:

" «*output* (a, b, c)» outputs via output device a the real value c with a format defined by the string b. The contents of the string (i.e. what appears between the string quotes) must have the syntactic form of an unsigned number (cf. 10.2), possibly preceded by a sign; furthermore, an arbitrary number of space symbols may be inserted at arbitrary places in this string. This string will serve as a model for the output of c: The space symbols, the period and the base ten of the string will appear at the respective places in the output of c (the space symbols will appear as blanks), while the sign of c will appear at the place of the sign in b but with the additional convention that a minus sign in b indicates that a positive sign of c will be suppressed in the output".

As an example, «*output* $(5, \text{'} \sqcup \sqcup -387.502 \sqcup 73_{10} + 17 \sqcup \sqcup \text{'}, 1.000)$» will produce the output $\boxed{ 100.000 \ 00_{10} - 02 }$ on the output device 5.

47.4. Economisation of ALGOL programs with aid of code procedures

If utmost efficiency is required, it may be felt that the translation of an ALGOL program into machine code should be done by an expert programmer rather than by an ALGOL compiler (at least as long as compilers cannot match the abilities of human programmers). In order to avoid forcing the programmer to translate the whole ALGOL program manually into machine code, one can single out those parts of the program in which most of the computing is done (e.g. the innermost loops), remove these parts from the program and declare them as code procedures. The other parts remain written in ALGOL, with appropriate calls of those code procedures being inserted between them. Of course this is not always feasible, but often a nearly optimal program can be obtained at the expense of optimizing only a small fraction of the whole program.

47.4.1. As an example, procedure *matinv* as given in 44.7.5 can be economized as follows:

```
«procedure matinv (n) trans: (a) exit: (fail) ;
     value n ;
     integer n ; array a ; label fail ;
     begin
        integer i, j, p ;
        procedure econom (n, p) trans: (a) ;          Code procedure for
           value n, p ;                               optimizing statement
           integer n, p ; array a ;                   delta in 44.7.5.
           code ;
        for p := 1 step 1 until n do
```

loop: **begin**
                if $a[p, p] = 0$ **then goto** *fail* ;
alpha: $a[p, p] := 1/a[p, p]$;
beta: **for** $j := 1$ **step** 1 **until** $p-1, p+1$ **step** 1 **until** n **do**
                            $a[p, j] := -a[p, j] \times a[p, p]$;
delta: $econom\,(n, p)\,trans:(a)$;
gamma: **for** $i := 1$ **step** 1 **until** $p-1, p+1$ **step** 1 **until** n **do**
                            $a[i, p] := a[i, p] \times a[p, p]$
             end *loop*
         end *matinv*»,

provided the body of *econom* is written in machine code as the equivalent of:

«**begin**
     integer i, j ;
     for $i := 1$ **step** 1 **until** $p-1, p+1$ **step** 1 **until** n **do**
         for $j := 1$ **step** 1 **until** $p-1, p+1$ **step** 1 **until** n **do**
             $a[i, j] := a[i, j] + a[i, p] \times a[p, j]$
end *econom*».

We observe that the economized program is scarcely any shorter than the original one, but it yields certainly a more efficient object program since all operations performed $O(n^3)$ times are expressed directly in machine code.

47.4.2. Let us now in the same way optimize the Banachiewicz process for solving linear systems (cf. 31.1): It is easy to see that it is statement *sum* which accounts for the total computing time being on the order of $O(n^3)$:

        «*sum*: **for** $j := 1$ **step** 1 **until** l **do** $t := t + a[i, j] \times a[j, k]$».

We may therefore realize a considerable saving in computing time by describing this operation (computation of the inner product $\sum_{j=1}^{n} a[i, j] \times a[j, k]$) as a procedure whose body is later transcribed into optimized machine code.

   However, if we try to compute this inner product by a call of procedure *inner* (declared in 44.1.3), we are faced with the problem that the formal operands x, y of *inner* are one-dimensional arrays while here we have to compute inner products of rows by columns of a two-dimensional array a. In ALGOL 58 we could have accomplished this in the most natural way with a call

            «*inner* $(l, a[i, \ ], a[\ , k])$ *res*: (t)»,

but this is not possible in ALGOL 60.

On the other hand, we could still handle this problem in full ALGOL 60 by means of the so-called *Jensen-device* (cf. § 52). However, since this solution, besides not being available in the SUBSET, is far from being optimal with respect to object program efficiency, we propose another solution. Indeed, since the inner product is such an important operation and yet is described by only a few ALGOL statements, it pays to construct a special inner product procedure for every situation. For our present purposes it is most appropriate to describe the computation of the inner product of the i-th row of a matrix $x[1:n, 1:l]$ with the k-th column of a matrix $y[1:l, 1:m]$ by a

«**real procedure** *rowcol* (n, m, l, i, k, x, y) ;
    value n, m, l, i, k ;
    integer n, m, l, i, k ; **array** x, y ;
    begin
       integer j ;
       real s ;
       $s := 0$;
       for $j := 1$ **step** 1 **until** l **do** $s := s + x[i, j] \times y[j, k]$;
       rowcol $:= s$
    end *rowcol*».

With this, the Banachiewicz process given in 31.1 can be rewritten as

«**begin**
    for $i := 1$ **step** 1 **until** n **do**
    begin
       for $k := 1$ **step** 1 **until** n **do**
       begin
          $l := $ **if** $i > k$ **then** $k - 1$ **else** $i - 1$;
          $t := a[i, k] + rowcol(n, n, l, i, k, a, a)$;
          $a[i, k] := $ **if** $k < i$ **then** $-t/a[k, k]$ **else** t ;
       end k ;
       $\vdots$
    ».

Now we can not only write the body of *rowcol* in machine code (thus producing a faster object program) but in doing so, we can perform the summation of the products just as well in double precision and in this way obtain more precise results.

§ 48. Parameter Procedures

Procedures which occur as formal or actual operands of other procedures are called *parameter procedures*. They allow quite unusual effects, which we are going to discuss in this section.

Examples of parameter procedures come mainly from two sources: Procedures that perform operations in which arbitrary functionals are involved, and procedures that require interference from outside during their execution.

48.1. Examples involving arbitrary functionals

48.1.1. Let us consider the summation of a finite series whose terms are arbitrary functions of a subscript k expressed by a real type function designator «$term(k)$». In order that arbitrary terms can be summed, the function procedure $term$ cannot be defined in the summation process but must be left open, i.e. quoted as formal operand:

«**real procedure** $sum(p, q, term)$;
   value p, q ;
   integer p, q ; **real procedure** $term$;
   comment calculates sum of $term(k)$ from $k=p$ through $k=q$;
   begin
      real s ;
      integer k ;
      $s := 0$;
      for $k := p$ **step** 1 **until** q **do** $s := s + term(k)$;
      $sum := s$
   end sum».

48.1.2. In order to use procedure sum for computing the inner product z of two vectors $a[1:n]$ and $b[1:n]$, the actual counterpart of $term$ in a call of sum must be declared such that the terms $a[k] \times b[k]$ are summed. In other words, we have to declare a function procedure, e.g. $scalp$, such that a corresponding function designator «$scalp(p)$» produces the value $a[p] \times b[p]$, and quote $scalp$ as actual counterpart of $term$ in a call of sum:

«**begin**
   comment quantities **array** a, b, **integer** n, **real** z, **real proce-**
           dure sum assumed global to this block ;
   real procedure $scalp(p)$;
   integer p ;
   comment $scalp$ uses global operands a, b ;
      $scalp := a[p] \times b[p]$;
   $z := sum(1, n, scalp)$;
end».

48.1.3. As a further example let us consider a procedure $euler$ for numerical integration of a system of differential equations

$$y_i' = f_i(x, y_1, y_2, \ldots, y_n) \quad (i = 1, 2, \ldots, n)$$

by Euler's method. One integration step is described by

«**begin**
    fct (x, y, n) *res*: (z) ;
    $x := x + h$;
    for $k := 1$ **step** 1 **until** n **do** $y[k] := y[k] + h \times z[k]$
end»,

where *fct* is a procedure which computes the quantities

$$z[k] = f_k(x, y[1], y[2], \ldots, y[n]) \quad (k = 1, 2, \ldots, n)$$

for given values of $n, x, y_1(x) = y[1], \ldots, y_n(x) = y[n]$. Thus *fct* defines the differential equations to be solved, but since *euler* should be able to integrate an arbitrary system of differential equations, *fct* cannot be declared within *euler* but must be quoted as formal operand:

«**procedure** *euler* (x, n, h, p, fct) *trans*: (y) *res*: (yy) ;
    value x, h, n, p ;
    real x, h ; **integer** n, p ; **array** y, yy ; **procedure** *fct* ;
    comment *euler* performs, for given initial values $x, y[1], y[2]$,
               $\ldots, y[n]$, p integration steps by Euler's method with
               increment h. The solution is obtained as an array
               $yy[1:p, 1:n]$, where $yy[j, k]$ denotes the value of the
               k-th component at the j-th meshpoint ;
  begin
    integer j, k ;
    array $z[1:n]$;
    for $j := 1$ **step** 1 **until** p **do**
    begin
        fct (x, y, n) *res*: (z) ;
        $x := x + h$;
        for $k := 1$ **step** 1 **until** n **do**
              $yy[j, k] := y[k] := y[k] + h \times z[k]$
    end j
  end *euler*».

48.1.4. The user of *euler* who wants to integrate a differential system must define the latter by declaring an appropriate procedure as actual counterpart of *fct* in a call of *euler*, e.g., if he wants to integrate the heat equation as indicated in 44.7.3:

«**begin**
    comment integer m, k, **array** $z1[1:m], z2[1:100, 1:m]$,
               procedure *euler* are assumed global to this block ;

procedure *equ* (*x*, *y*, *n*) *res*: (*z*) ;
   value *x*, *n* ;
   real *x* ; **integer** *n* ; **array** *y*, *z* ;
   begin
                  procedure body as in 44.7.3.
   end *equ* ;
   for *k* := 1 **step** 1 **until** *m* **do** *z1* [*k*] := 0 ;
   euler (0, *m*, 0.01, 100, *equ*) *trans*: (*z1*) *res*: (*z2*) ;
end ».

After termination of this block the desired solution is found in the array *z2*, the *j*-th row of which contains the heat distribution at the time $t = h \times j$.

48.2. Execution of parameter procedures

48.2.1. The precise effect of a procedure call involving parameter procedures can always be analyzed by iterated application of the substitution rule, beginning with the outermost procedure call and ending with the innermost.

For instance, if we have the situation:

«**begin**
   procedure *zzz* (*a*, *b*, *c*, *d*) ; **real** *a*, *b*, *c* ; **label** *d* ; } Declaration
      if *b* > 0 **then goto** *d* **else** *c* := *a*/*b* ;   } for *zzz*.
   xxx (*f*, *zzz*)
end »,

where procedure *xxx* is declared in some outer block as:

«**procedure** *xxx* (*w*, *yyy*) ;
      real *w* ; **procedure** *yyy* ;
      begin
         real *u*, *v* ;
            ⋮
   13:   *yyy* (*u*, *v*, *w*, *label*) ;
   label:
      end *xxx* »,

then the rules of 45.2 are applied first to the call «*xxx* (*f*, *zzz*)». Hereby this call is replaced by the equivalence block, and the call «*yyy* (*u*, *v*, *w*, *label*)» occurring in the latter is modified into a call «*zzz* (*u*, *v*, *f*, *label*)», to which the rules of 45.2 are applied again.

48.2.2. *The semistatic rule.* Despite the fact that the analytical method of 48.2.1 above is the only legal one for defining the effect of procedure calls involving parameter procedures, computing practice requires a slightly different approach. Indeed, a parameter procedure serves essentially to allow for modifications of a procedure at call time; what we need to know is therefore the operational behavior of the modified procedure, more precisely, the behavior of a procedure X under consideration of a specific procedure Z given as actual counterpart of a formal operand Y (of X). For the user of X it is desirable to have the properties of the modified X described again as a procedure $X1$, which can be constructed as follows:

The rules of 45.2 are applied to the calls of Y occurring inside X, only that the body of Z is used in place of the (nonexistent) body of Y. After that, Y is removed from the formal parameter part and specification part of procedure X.

For the example mentioned in 48.2.1 above this comes as follows:

«**procedure** $xxx1\,(w)$;
    real w ;
    begin
      real u, v ;
      $\vdots$

$l3:$ **begin**
        real $æ$;
        if $v>0$ **then goto** $label$ **else** $w := u/v$ } Equivalence block for call of zzz.
      end zzz ;
$label:$
    end $xxx1$».

Now the call «$xxx\,(f, zzz)$» has the same effect as the call «$xxx1\,(f)$» of the new procedure; for the latter the rules of 45.2 yield the same equivalence block as would have resulted through the orthodox rules mentioned in 48.2.1:

«**begin**
    real u, v ;
    $\vdots$

$l3:$ **begin**
      real $æ$;
      if $v>0$ **then goto** $label$ **else** $f := u/v$ { f is substituted for w
    end zzz ;
$label:$
 end $equivalence\ block$».

Applying the semistatic rule to the combination $euler/equ$ as declared in 48.1.3 and 44.7.3, we obtain:

«**procedure** $euler1(x, n, h, p)$ trans: (y) res: (yy) ;
    value x, n, h, p ;
    real x, h ; **integer** n, p ; **array** y, yy ;
    comment integrates a specific differential equation by Euler's
              method ;
  begin
    integer j, k ;
    array $z[1:n]$;
    for $j := 1$ **step** 1 **until** p **do**
    begin
      begin
        integer k ;
        $z[1] := -2 \times y[1] + y[2]$;
        for $k := 2$ **step** 1 **until** $n-1$ **do**
          $z[k] := y[k-1] - 2 \times y[k] + y[k+1]$;
        $z[n] := y[n-1] - 2 \times y[n] + t \times x$
      end equ ;
      $x := x + h$;
      for $k := 1$ **step** 1 **until** n **do**
        $yy[j, k] := y[k] := y[k] + h \times z[k]$
    end j
  end $euler1$».

(Body of procedure equ.)

Note that this new procedure *euler1* has a global operand t which was introduced through procedure *equ*. According to the environment rule for global parameters, this t refers therefore to a quantity valid in the environment of the declaration for *equ*; in fact, t links *euler1* to this environment.

48.2.3. Provided procedure X (with procedure Y as formal operand) is independent, the execution of a call of X with Z as actual counterpart of Y can be visualized as follows:

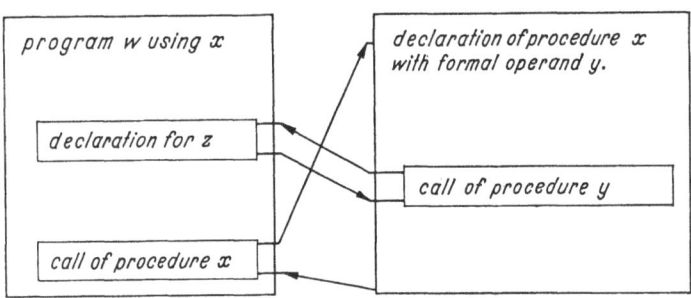

Fig. 40

As this picture indicates, it is as if upon call of X a jump from the "program" W to the declaration for X occurred, and upon call of Y (inside X) a jump back to the declaration of Z, which is located in W (this is what we call an *intermediate return* to the program which called X). After execution of Z, execution continues in X, and after termination of X, a jump back to W occurs.

48.3. Interference with the execution of a procedure[1]

48.3.1. Procedures which completely define the operations to be performed upon a call are sometimes too inflexible for practical purposes. For instance, the proper termination of an iterative process may require information which is simply not available to the designer of the procedure, while the user may know more about it. In other cases the convergence could be improved if only the user of the procedure could give some clues concerning the distribution of the eigenvalues of a certain matrix.

In order to enable the user of a procedure X to apply his knowledge for influencing the execution of the procedure, the designer of X selects certain *breakpoints* in the body of X where he thinks the user might wish to intervene. At these points he then places calls of an unspecified procedure Y and chooses as actual operands of this call all those internal quantities of X that might be used for influencing the execution of X, plus one extra operand for enumerating the various calls of Y. Y is then quoted as formal operand of X.

On the whole, the declaration for X will obtain the following structure:

«**procedure** $X(\ldots, Y)$;
   $\ldots$; **procedure** Y ;
   begin
      real a ; **integer** b ; **switch** c ;
      $\vdots$
      $Y(1, a, b, c, \ldots)$;
      $\vdots$
      $Y(2, a, b, c, \ldots)$;
      $\vdots$
      $Y(3, a, b, c, \ldots)$;
      $\vdots$
   end X».

In certain cases, several such parameter procedures $Y_1, Y_2, \ldots,$ may be needed, while in other cases Y is called only once, making the enumeration parameter superfluous.

---

[1] Compare also [30].

48.3.2. Consider, as an example, the orthogonalisation procedure as given by SCHWARZ [*34*], §8 (slightly modified), in which we insert two calls of a parameter procedure *cont*:

```
«procedure orth (n, p, cont) trans: (a) res: (r) exit: (zero) ;
    value n, p ;
    integer n, p ;  array a, r ;  label zero ;  procedure cont ;
    begin
        real u ;
        integer i, j, k, s ;
        for k := 1 step 1 until p do
loop:   begin
            for j := 1 step 1 until p do r [j, k] := 0 ;      ⎰First call of parame-
            cont (1, k, p, n, a, r, exit) ;                   ⎱ter procedure cont.
enter:      for j := 1 step 1 until k − 1 do
            begin
                u := 0 ;                                       ⎫
                for i := 1 step 1 until n do                   │ Make k-th column of
                    u := u + a[i, j] × a[i, k] ;               │ A orthogonal to all
                for s := 1 step 1 until n do                   ⎬ previous columns
                    a[s, k] := a[s, k] − u × a[s, j] ;         │ and build up the co-
                r[j, k] := r[j, k] + u ;                       │ efficients r [j, k].
            end j ;                                            ⎭
exit:       u := 0 ;
            for i := 1 step 1 until n do                       ⎫ Determine length of
                u := u + a[i, k]↑2 ;                           ⎬ orthogonalized
            r[k, k] := sqrt (u) ;                              ⎭ k-th column.

            cont (2, k, p, n, a, r, enter) ;                   ⎰Second call of pa-
                                                               ⎱rameter procedure cont.

            if r[k, k] = 0 then goto zero ;
            for s := 1 step 1 until n do                       ⎰ Normalize
                a[s, k] := a[s, k]/r[k, k] ;                   ⎱ k-th column.
        end k ;
    end orth».
```

In this procedure the columns of the matrix A given as an array $a[1:n, 1:p]$ are orthonormalized, i.e. A is decomposed into $U \times R$ such that after termination the array $a[1:n, 1:p]$ contains the elements of U (which has orthonormal columns) while the resulting array $r[1:p, 1:p]$ contains the elements of the upper triangular matrix R.

The parameter procedure *cont* can be used in various ways; for instance, we could declare as actual counterpart of *cont*:

«**procedure** *c1* (*nr, k, p, n, a, r, label*) ;
    value *nr* ;
    integer *k, p, n, nr* ; **array** *a, r* ; **label** *label* ;
    if *nr* = 1 **then goto** *label*».

This would have the effect — as becomes obvious through the semistatic rule — that the columns of A are merely normalized without changing their direction, since the whole orthogonalisation part (statement *enter*) is skipped.

48.3.3. However, if we declare

«**procedure** *c2* (*nr, k, p, n, a, r, label*) ;
    value *nr* ;
    integer *nr, k, p, n* ; **array** *a, r* ; **label** *label* ;
    begin
      if *nr* = 1 **then** *flag* := **false** ;
      if $nr = 2 \wedge \neg flag \wedge r[k, k] \neq 0$ **then**
      begin
        real *f* ;
        integer *l* ;
        flag := **true** ;

        f := 0 ;
        for *l* := 1 **step** 1 **until** *k*—1 **do**
            $f := f + r[l, k]\uparrow 2$;
        if $f > 100 \times r[k, k]\uparrow 2$ **then goto** *label*;

> Go to repetition of orthogonalisation if $r[k, k]\uparrow 2$ becomes small compared to $r[1, k]\uparrow 2 + r[2, k]\uparrow 2 + \cdots + r[k-1, k]\uparrow 2$, which means that the length of the k-th column has collapsed.

      end *if nr* = 2
    end *c2*»

and use this as actual counterpart of *cont* in a call of *orth*, it helps to avoid deterioration of the orthogonality of the columns of U, which otherwise might take place in certain cases. Indeed, according to the semistatic rule, and with some obvious simplifications, procedure *orth* becomes equivalent to:

«**procedure** *orth1* (*n, p*) *trans*: (*a*) *res*: (*r*) *exit*: (*zero*) ;
    value *n, p* ;
    integer *n, p* ; **array** *a, r* ; **label** *zero* ;
    comment global operand: *flag* ;

```
       begin
          real u ;
          integer i, j, k, s ;
          for k := 1 step 1 until p do
loop:     begin
             for j := 1 step 1 until p do r[j, k] := 0 ;

             flag := false ;

enter:       for j := 1 step 1 until k−1 do
             begin
                u := 0 ;
                for i := 1 step 1 until n do
                          u := u+a[i, j] × a[i, k] ;
                for s := 1 step 1 until n do
                          a[s, k] := a[s, k] − u × a[s, j] ;
                r[j, k] := r[j, k] + u
             end j ;
exit:        u := 0 ;
             for i := 1 step 1 until n do u := u+a[i, k]↑2 ;
             r[k, k] := sqrt(u) ;
             if ¬flag ∧ r[k, k] ≠ 0 then
             begin
                real f ;
                integer l ;
                f := 0 ;
                flag := true ;
                for l := 1 step 1 until k−1 do
                          f := f+r[l, k]↑2 ;
                if f > 100 × r[k, k]↑2 then goto enter ;
             end if ¬flag ;
             if r[k, k] = 0 then goto zero ;
             for s := 1 step 1 until n do
                          a[s, k] := a[s, k]/r[k, k]
          end k ;
       end orth1».
```

Marginal notes:

for the block beginning `flag := false ;` —
> Equivalent of first call of *cont* with *c2* as actual counterpart of *cont*.

for the block beginning after `r[k, k] := sqrt(u)` —
> Equivalent of second call of *cont* with *c2* as actual counterpart of *cont*.

Quite obviously this modified procedure *orth1* repeats the orthogonalisation of the k-th column of A in case its length has collapsed during the first orthogonalisation, and thus improves the orthogonality of the columns of the resulting matrix U.

48.3.4. *An unusual application.* Parameter procedures may often be used to change completely the functional behavior of a given procedure; whether this is useful or not is of course another question.

Let us again consider procedure *orth* as declared above in 48.3.2. From this declaration we see that the components of the array a, which appears as the fourth formal operand, need not be given at call time but can be brought into the computation via the parameter procedure *cont*[1]. More precisely, at the latest for the beginning of the k-th turn of the k-loop (statement *loop*) the k-th column of the array a must be available. This makes it possible to construct the k-th column as follows: After columns 1 through $k-1$ of the matrix A have been orthonormalized and thus the column vectors $\vec{u}_1, \vec{u}_2, \dots, \vec{u}_{k-1}$ of the matrix U have been computed, the k-th column vector $\vec{a}_k$ of A is produced as the product of a given matrix $B = b[1:n, 1:n]$ with $\vec{u}_{k-1}$. With this $\vec{a}_k$, statement *loop* will produce the orthonormalized vector $\vec{u}_k$ and the k-th column of the matrix R such that the following relation holds:

$$\vec{a}_k = B\,\vec{u}_{k-1} = r_{1,k}\,\vec{u}_1 + r_{2,k}\,\vec{u}_2 + \cdots + r_{k-1,k}\,\vec{u}_{k-1} + r_{kk}\,\vec{u}_k.$$

If we set out with an arbitrary vector $\vec{a}_1 = \vec{w}$ and proceed for $k=2$, ..., $n+1$, it turns out that we perform exactly the *Arnoldi method* [3] for transforming an arbitrary matrix B to Hessenberg form:

$$B \to R_1 = U^T B U,$$

where U is the orthogonal matrix with columns $\vec{u}_k$, and

$$R_1 = \begin{cases} r_{12} & r_{13} & \cdots\cdots & r_{1,n+1} \\ r_{22} & r_{23} & & \vdots \\ 0 & r_{33} & & \vdots \\ 0 & 0 & \ddots & \vdots \\ \vdots & \vdots & & \vdots \\ 0 & 0 & r_{nn} & r_{n,n+1} \end{cases}$$

contains the 2-nd through $n+1$-st columns of R. To achieve this effect, we declare

```
«procedure arnold (nr, k, p, n, a, r, label) ;
     value nr ;
     integer nr, k, p, n ; array a, r ; label label ;
     comment global operands: b[1:n, 1:n], w[1:n] ;
     if nr = 1 then
```

---

[1] This is only possible for operands called by name since for operands called by value, rule 45.2.1 would put a meaningless assignment statement into the floor of the equivalence block.

begin
  integer i, j ;
  real s ;
  if $k = 1$ **then for** $i := 1$ **step** 1 **until** n **do** $a[i, k] := w[i]$;
  if $k \neq 1$ **then**
    for $i := 1$ **step** 1 **until** n **do**
    begin
      $s := 0$;
      for $j := 1$ **step** 1 **until** n **do**
                $s := s + b[i, j] \times a[j, k-1]$;
      $a[i, k] := s$
    end
  end *arnold* »

and then call *orth* by

    «$orth(n, n+1, arnold)$ *trans*: (a) *res*: (r) *exit*: $(zero)$».

This statement requires that the value of n be given and that the arrays a, r be declared (at least) as

    «**array** $a[1:n, 1:n+1], r[1:n+1, 1:n+1]$»,

while the arrays $b[1:n, 1:n]$ and $w[1:n]$ must be declared *and* have values. After termination of this procedure statement, U is the matrix contained in the columns 1 through n of array a, while R_1 is found in columns 2 through $n+1$ of array r.

48.3.5. *Nested parameter procedures.* While the above problem is of the simple structure indicated in Fig. 40 (48.2.3), still more complicated situations may occur in practice.

For instance, we may desire to perform experiments with various numerical methods for integrating differential equations. We assume these methods to be described formally as

    «**procedure** *method* (x, n, h, p, fct) *trans*: (y) *res*: (yy) ; ...»

with the same meaning of the formal operands as in procedure *euler* (cf. 48.1.3).

In order that these methods can be compared, they are all applied to the same type of differential equation

$$(j(x) y'')'' = m(x) y,$$

which is integrated in p equal steps from 0 to 1. To this end, auxiliary variables $y_1 = y$, $y_2 = y'$, $y_3 = j(x) y''$, and $y_4 = (j(x) y'')'$ are introduced.

To perform this task such that p, the positive functions $j(x)$ and $m(x)$ as well as the initial values $w[k] = y_k(0)$ $(k = 1, 2, 3, 4)$ can be varied, a procedure *metex* is declared as follows:

«**procedure** *metex* ($p, j, m,$ *method*) *trans*:(w) *res*:(yy) ;
    value p ;
    integer p ; **array** w, yy ; **procedure** *method* ;
    real procedure j, m ;
    begin
        array $y, z[1:4]$;
        procedure *sys* (x, y, n) *res*:(z) ; **value** n, x ;
            real x ; **integer** n ; **array** y, z ;
            begin
                $z[1] := y[2]$;
                $z[2] := y[3]/j(x)$;
                $z[3] := y[4]$;
                $z[4] := m(x) \times y[1]$
            end *sys* ;

       method $(0, 4, 1/p, p, sys)$ *trans*:(w) *res*:(yy) ;
end *metex*».

> Procedure *sys* to be used as actual counterpart of *fct* in a call of the formal operand *method*, where the latter represents the method to be tested.

> Call of the formal operand *method*.

The effect of a call «*metex* $(p, jj, mm,$ *euler*) *trans*:(w) *res*:(yy)» of this procedure might be pictured as follows:

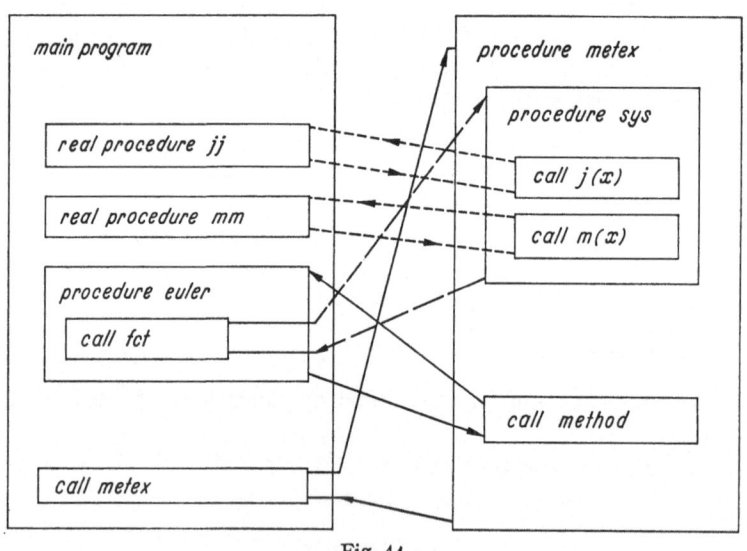

Fig. 41

48.4. Some programming problems

48.4.1. Usually a procedure P is declared first and only later is it used, whereby the user may infer from the corresponding declaration what

a call of the procedure should look like. For parameter procedures, however, it is just the other way round: All that the user of procedure *orth*
(to give an example) can see of the parameter procedure *cont* are two calls

«*cont* $(1, k, p, n, a, r, exit)$» and «*cont* $(2, k, p, n, a, r, enter)$»,

from which he must deduce how the corresponding actual procedure
should be declared in order that the above calls meet the rules given
in § 45 (in case of a formal function procedure also § 46 must be observed).
The necessary properties of the operands of *cont* can be found only by
inspection of procedure *orth*.

In our example, we quickly see that the seven operands are (from
left to right):

1)   an integer variable (which must be called by value in order to
       allow for a numerical constant as corresponding actual operand),
2) 3) 4) three integer variables,
5) 6)   two real arrays,
7)   a label,

but in other cases the analysis may be tedious and in certain cases
(cf. 45.4.4) the types of the operands of a parameter procedure are even
indeterminate.

48.4.2. *Global operands of parameter procedures.* If a differential equation
containing arbitrary coefficients, e.g.

$$y'' + k \times (y')\uparrow 3 + l \times y = 0,$$

should be integrated with the aid of procedure *euler*, such coefficients
can enter the computation only as global operands of the actual counterpart of *fct* because the rules of ALGOL simply leave no other choice.
We therefore declare:

«**procedure** *fctact* (x, y, n) *res*: (z) ;
  real x ; **integer** n ; **array** y, z ;
  comment global operands: k, l ;
  begin
    $z[1] := y[2]$;
    $z[2] := -k \times y[2]\uparrow 3 - l \times y[1]$
  end».

This is also quite natural, since on one hand the coefficients k, l are
alien to procedure *euler*, while on the other hand the environment rule
(cf. 44.3.2) clearly states that the quantities k, l automatically belong
to the outside of *euler* and are considered different from local quantities
of *euler* that may have the same names. In other words, if Z is a procedure

to be used as actual counterpart of a procedure Y which is formal operand of X, the user of X may choose the names of global operands of Z without worrying about names of local quantities of X since a conflict with these names is impossible.

48.4.3. *Access to internal quantities of* X. An adverse effect of the otherwise very useful environment rule is that a parameter procedure Y of a procedure X allows access only to those internal quantities of X which are quoted as actual operands in at least one call of Y.

As an example, the variable u, which is local to the body of procedure *orth* (cf. 48.3.2), is not quoted as actual operand in a call of *cont*. Declaring the actual counterpart of *cont* as

«**procedure** $c3(nr, k, p, n, a, r, label)$;
    value nr ;
    integer nr, k, p, n ; **array** a, r ; **label** $label$;
      if $nr=2$ **then begin** $line$; $print(k)$; $print(u)$ **end**»[1]

would therefore not have the effect of printing $u = \sum_{i=1}^{n} a[i, k] \uparrow 2$, as it might be hoped, but would instead print the value of a variable u existing outside *orth*, or produce an undefined effect.

Likewise, we could not enforce a jump to the label *loop* via the parameter procedure *cont* since *loop* is not quoted as actual parameter of *cont*.

---

[1] For the meaning of procedures *line, print* see 43.2.

Chapter VIII

Input and Output

No ALGOL program is complete without providing for transfers of initial data and final results from and to the outside world. In ALGOL such transfers may be done either by code procedures which have been designed for that purpose (see, for instance, the so-called *Knuth report* [22]), or through the standard I/O-procedures *insymbol, outsymbol, inreal, outreal, inarray* and *outarray*. These latter we are now going to describe.

§ 49. The Standard I/O-Procedures of ALGOL

At a meeting held in March 1964 at Tutzing, Bavaria, the IFIP Working Group on ALGOL (WG 2.1) decided that six standard procedures for describing input- and output-operations in an abstract manner (i.e. without reference to specific devices) should be added as fixed constituents of the language[1]. In the future, therefore, these procedures may, like the standard functions *sin, cos*, etc., be called in an ALGOL program without being declared there. In fact, it is as if corresponding declarations had been given in a block embracing all ALGOL programs.

49.1. Syntax

Although the declarations for the six standard I/O-procedures never do actually appear in an ALGOL program, we can, nevertheless, exhibit these declarations in order to specify the conditions which must be met by corresponding calls[2]:

«**procedure** *insymbol* (*a, b, c*) ;
  value *a* ;
  integer *a, c* ; **string** *b* ;
  code».

«**procedure** *outsymbol* (*a, b, c*) ;
  value *a, c* ;
  integer *a, c* ; **string** *b* ;
  code».

---

[1] See [21]. Meanwhile this decision has been confirmed by the IFIP council (meeting held at Prague, May 16, 1964).

[2] It should be recognized that in the SUBSET the names of these six procedures are equivalent to *insymb, outsym, inreal, outrea, inarra, outarr*, respectively, which therefore should be considered as reserved names.

«**procedure** *inreal* (*a*, *b*) ;
    value *a* ;
    real *b* ; **integer** *a* ;
    code».

«**procedure** *outreal* (*a*, *b*) ;
    value *a*, *b* ;
    real *b* ; **integer** *a* ;
    code».

«**procedure** *inarray* (*a*, *b*) ;
    value *a* ;
    integer *a* ; **array** *b* ;
    code».

«**procedure** *outarray* (*a*, *b*) ;
    value *a* ;
    integer *a* ; **array** *b* ;
    code».

Here the symbols «**code**» denote the respective procedure bodies, which of course cannot be described directly in terms of ALGOL but are primitives of the ALGOL language itself.

49.2. Semantics

The precise meanings of these standard I/O-procedures are defined by describing the effect of corresponding calls:

49.2.1. A call «*insymbol* (*a*, *b*, *c*)» reads the next symbol from an input medium designated by the value of *a* and compares this symbol to the sequence of basic symbols contained in the string *b*. If the symbol matches the *k*-th symbol in the string *b* (counting from left to right, beginning with 1), then the value *k* is assigned to the variable *c*. However, if the symbol read from the medium does not match any of the symbols contained in the string *b*, then the value 0 is assigned to *c* (see, however, 49.4.2).

49.2.2. «*outsymbol* (*a*, *b*, *c*)» records the *c*-th basic symbol contained in the string *b* (counting as in 49.2.1) on an external medium which is determined by the current value of *a* (see, however, 49.4.2).

49.2.3. «*inreal* (*a*, *b*)» reads the next following number from an external medium designated by the current value of *a* and assigns this number to the variable *b*.

49.2.4. «*outreal* (*a*, *b*)» records the value of *b* on an external medium which is selected according to the current value of *a*.

49.2.5. «*inarray* (*a*, *b*)» reads real numbers one by one from an external medium defined by the value of *a* and assigns these numbers to the components of the array *b* (Exactly as many numbers are read from the external medium as are needed to fill the array *b*).

49.2.6. «*outarray* (*a*, *b*)» records all components of the array *b* on an external medium designated by the value of *a*.

49.2.7. It is explicitly understood that reading by *inreal* and recording by *outreal* are compatible operations such that e.g. a number recorded via *outreal* can be read again through *inreal* (the same is true for the pairs *insymbol*/*outsymbol* and *inarray*/*outarray*). Furthermore, it is assumed that *inreal* and *outreal* are also compatible with *inarray* and *outarray* in the sense that e.g. numbers recorded via *outreal* can be read again by *inarray*.

49.3. Further remarks

49.3.1. Since real type numbers and variables are involved in the procedures *inreal*, *outreal*, *inarray*, *outarray*, the remarks concerning computer limitations (cf. 8.2) apply also here.

49.3.2. For *inarray* and *outarray* the order in which the components of the array *b* are transferred is defined to be what for matrices (two-dimensional arrays) is usually called "row-wise". More precisely: $b[i_1, i_2, \ldots, i_p]$ is transferred *before* $b[j_1, j_2, \ldots, j_p]$ provided we have for some $h \le p$:

$$i_l = j_l \quad \text{for} \quad l = 1, 2, \ldots, h-1, \quad \text{but} \quad i_h < j_h.$$

Moreover, these procedures always transfer *all* components of the array appearing as the second actual operand.

As a consequence, a call «*outarray* (15, *p*)», where *p* is declared e.g. as «**array** $p[-4:5, 1:50, 0:20]$», is equivalent to

«**for** $j1 := -4$ **step** 1 **until** 5 **do**
    for $j2 := 1$ **step** 1 **until** 50 **do**
        for $j3 := 0$ **step** 1 **until** 20 **do** *outreal* $(15, p[j1, j2, j3])$».

49.3.3. It is explicitly understood that calls of the six standard I/O-procedures automatically include the movement of the external medium in order to make it ready for input or output of the next item. Of course, where the external medium is for instance a magnetic drum, the movement means simply stepping a counter.

49.3.4. The *channel number* appearing as the first parameter in all standard I/O-procedures should be understood in the most general sense. In actual computing it may mean a paper tape punching or reading station of the computer, or an addressed section of a magnetic tape or disc file (in the latter case the address is part of the channel number).

49.4. Control operations

49.4.1. *The IFIP convention.* Through one of the examples given in the IFIP report [21], the latter implicitly states that if input is done via a channel which previously had been used for output, the channel is reset into a position following that data which had been read most recently through that channel. Likewise, if we switch from input to output on the same channel, the new output is positioned after the most recent output. Thus a sequence of input- and output-operations on the same channel, e.g.

out, out, out, in, in, out, out, out, in, in, out, out,
out, in, in, in, in, out, in, in,

will initiate the following movements of the reading and recording mechanism associated with this channel:

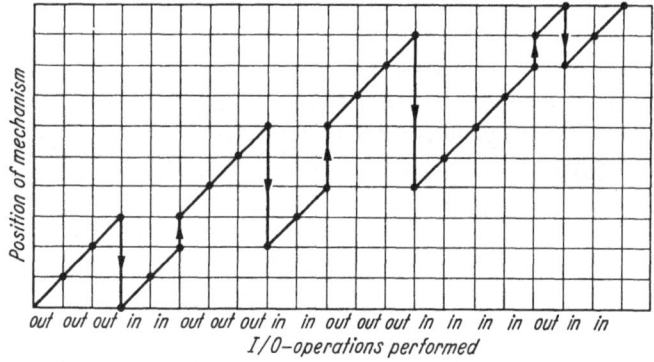

Fig. 42

Since after this, all recorded items have been read again, an output operation must now follow, otherwise the further course of the process would be undefined.

By this convention it is obviously impossible to read a recorded item twice, i.e. it is as if the information on the external medium were destroyed upon reading. If repeated reading is desired, we have to adhere to the convention of 49.4.2.

49.4.2. *Explicit resetting of a channel,* but also operations such as carriage return and line feed of a console typewriter, can be performed with the aid of procedure *outsymbol* just by giving the third actual parameter a negative value. As an example,

$$«outsymbol\,(15, 'rewind\ tape\ 15', -13)»^{[1]}$$

may cause rewinding of tape 15. Similarly, setting of an "end of file" mark may be done in this way. The IFIP resolution explicitly permits such operations but does not prescribe any details, which therefore are entirely in the hands of the implementors.

However, for the purposes of the present chapter the following convention on control operations is adopted:

Value of the third parameter c of *outsymbol*	Control operation associated with the value c
-1	Setting a mark "end of record" (For a printer: new line)
-2	Setting a mark "end of file" (For a printer: new page)
-3	Setting a mark "third order end"
$\vdots$	$\vdots$
-11	Reset channel to most recent mark "end of record"
-12	Reset channel to most recent mark "end of file"
-13	Reset channel to most recent mark "third order end"
$\vdots$	$\vdots$

It is assumed that "end of record" and "end of file" marks are skipped on all input operations except in calls of *insymbol*, in which case the corresponding negative values are produced and assigned to the variable appearing as the third actual operand. It is furthermore assumed that resetting a channel, e.g. by -11, puts the corresponding mechanism into a position immediately in front of the respective mark, in order that a further reset operation spaces the mechanism back by another record. Finally, where no mark corresponding to a reset operation can be found, the whole channel is reset.

49.4.3. *Combinations of explicit and implicit resetting.* After resetting a channel explicitly, e.g. by «*outsymbol* $(ch, '|', -12)$», we may later use that channel according to the IFIP convention. However, where a channel is reset explicitly as above by «*outsymbol* $(ch, '|', -12)$» after

---

[1] If c is negative, the contents of the string appearing as second actual operand are irrelevant and have the same effect as a comment.

having been used with implicit resetting (i.e., under the IFIP convention), all information previously recorded on this channel following the
corresponding "end of file" mark must be considered as having been
destroyed.

49.5. The I/O-procedures of § 43

The previously used I/O-procedures *line, print, read, prtext* can now
be expressed in terms of the standard I/O-procedures: Assuming that
channel 2 is associated permanently with a printer, channel 1 with a
paper tape reader, we obtain:

«**procedure** *line* ; *outsymbol* (2, 'carriage return', − 1) »,

«**procedure** *print* (*x*) ;
   value *x* ;
   real *x* ;
   outreal (2, *x*) »,

«**procedure** *read* (*x*) ;
   real *x* ;
   inreal (1, *x*) »,

«**procedure** *prtext* (*s*) ;
   string *s* ;
   begin
      integer *k*, *l* ;
      l := *length* (*s*) ; **comment** *length* is a standard function ;
      for *k* := 1 **step** 1 **until** *l* **do** *outsymbol* (2, *s*, *k*)
   end *prtext* ».

It is probably worthwhile to mention that in the last example
«*length* (*s*)» computes the length of the formal string *s* (i.e. the number
of symbols contained in this string), whereupon *outsymbol* is called once
for every *k* = 1, 2, ..., *l*, producing for every *k* the *k*-th basic symbol
of *s* as output. Thus on the whole, string *s* is written by the printer
associated with channel 2.

§ 50. Applications of Procedures *insymbol, outsymbol*

Since the procedures *insymbol* and *outsymbol* give access to single
characters on the external medium, it is obvious that they can be used
to perform all sorts of data processing operations.

50.1. Input and output of pseudostrings

If *a* [0:*bigm*] is a pseudostring according to the conventions of § 37,
it can be output by the following statement:

«**for** $k := 1$ **step** 1 **until** $a[0]$ **do**
    $outsymbol\,(channel,\,`012\ldots XYZ',\,a[k])$ »,

in which '$012\ldots XYZ$' is supposed to be the string of all basic symbols of ALGOL in the same order as shown by the correspondence table in 37.1. Indeed, according to the definition of *outsymbol*, this piece of program picks up the components of the pseudostring a one by one and if e.g. $a[k]=j$, searches for the j-th basic symbol in the string '$012 \ldots XYZ$' and records it via channel *channel*. Thus, obviously, the pseudostring a is output as an actual string of symbols.

In order that the lengthy string '$012\ldots XYZ$' of 116 symbols need not be written anew for every output operation, a procedure *pstout* for performing output of a pseudostring, including an "end of record" mark at its end, may be written as follows:

«**procedure** *pstout* (*channel*, a) ;
    value *channel* ;
    integer *channel* ; **integer array** a ;
    begin
        integer k ;
        for $k := 1$ **step** 1 **until** $a[0]$ **do**
                $outsymbol\,(channel,\,`012\ldots XYZ',\,a[k])$;
        $outsymbol\,(channel,\,`end\ of\ record',\,-1)$
    end *pstout*».

Input is somewhat more complicated, since we cannot assume that on the input medium the number of characters is given in front of the data. Instead, we require that the data be terminated by an "end of record" mark yielding upon input the integer value -1 (cf. 49.4.2). With this convention we obtain:

«**procedure** *pstin* (*channel*, a) ;
    value *channel* ;
    integer *channel* ; **integer array** a ;
    begin
        integer k, *aux* ;
        $k := 0$;
  ziv: $k := k+1$;
        $insymbol\,(channel,\,`012\ldots XYZ',\,aux)$;
        if $aux = -1$ **then** $a[0] := k-1$
        else
            begin $a[k] := aux$; **goto** *ziv* **end**
    end *pstin*».

50.2. Punched card reading

For reading from and recording on discontinuous media such as punched cards, a slightly different approach is more appropriate. Indeed, a punched card is in a certain sense one unit of information, and therefore the card structure should not be lost automatically upon input into the computer. To this end we insert, after reading of one card, an "end of record" symbol (-1 in our representation) into the generated pseudostring. At the same time we have to observe that only the first p columns of one card contain relevant information, while columns $p+1$ through 80 are used for identification (usually $p=72$, but we leave p variable).

In the following we make the assumption that a call «*outsymbol* (cr, '␣', -1)» causes the card reader designated by the current value of cr to feed one card, and that the subsequent calls of *insymbol* read the columns of the new card one by one from 1 through 80 (If we attempt to read more than 80 columns, the further action of the program is undefined).

Under these hypotheses the following procedure *incard* reads a deck of cards and places the data into the pseudostring $a[0:bigm]$ occurring as the fourth formal operand. The reading is terminated as soon as an "end of file" mark is found on one card or if the pseudostring is exhausted, in which case a jump to the formal label *full* occurs. In both these cases the last card is ejected and no further action is taken[1]:

«**procedure** *incard* (cr, $bigm$, p) res: (a) $exit$: ($full$) ;
    value cr, $bigm$, p ;
    integer cr, $bigm$, p ; **integer array** a ; **label** $full$;
    begin
        integer j, k, l, aux ;
        for $k:=0$, l **while** $l<bigm$ **do**
        begin

	for $j:=k+1$ **step** 1 **until** $k+p$ **do**	
	begin	⎫
	insymbol (cr, '012 ... XYZ', aux) ;	Read one card, k denoting position of the
$s1$:	$a[j] := aux$;	last element of the
	$l:=j$;	previous card in the
	if $j=bigm \vee aux=-2$ **then goto** $finis$	array a.
	end j ;	⎭
	$l:=l+1$;	⎫ Manipula-
$s2$:	$a[l]:=-1$;	tions after
	if $l \neq bigm$ **then** *outsymbol* (cr, '$cardfeed$', -1)	reading full
	end k ;	⎭ card.

---

[1] Here again '012 ... XYZ' stands for the string of all basic symbols of ALGOL.

finis: $a[0] := l$;
      outsymbol (*cr*, '*eject last card*', -1) ;
      if $aux \neq -2$ **then goto** *full*
   end *incard*».

If it is desired to store the information in packed form (cf. 37.5), we need only replace the statements *s1* and *s2* in the above program by

$$«pack (aux, j, a)» \quad \text{and} \quad «pack (-1, l, a)»$$

respectively, assuming that the packing includes transformation of negative integers into complementary form, e.g. -1 into 127.

50.3. Simulation of an output buffer

Let us assume that an output buffer region has been reserved by declaring a two-dimensional array *buffer* at the beginning of the whole program:

«**integer array** $buffer[1:25, 0:120]$»[1].

It is assumed that every row of this array, i.e. the components $buffer[i, 1]$ through $buffer[i, 120]$, contains one record whose actual length is given by the component $buffer[i, 0]$.

50.3.1. To output rows a through b of this buffer via channel ch, the following procedure may be declared:

```
«procedure outbuf (ch, a, b) ;
    value ch, a, b ;
    integer ch, a, b ;
    comment uses global operand: integer array buffer ;
    begin
        integer i, j ;
        for i := a step 1 until b do
        begin
            comment output of one line of the buffer ;
            for j := 1 step 1 until buffer [i, 0] do
                    outsymbol (ch, '012 ... XYZ', buffer [i, j]) ;
            outsymbol (ch, 'end of record', −1) ;
blanks:     for j := 1 step 1 until 120 do buffer [i, j] := 42
        end i
    end outbuf».
```

---

[1] It is agreed that usually this extra storage requirement of 3025 storage positions cannot be afforded. However, it is assumed that in practice the data would be packed. What we show here must therefore be understood as a model for a more efficient setup.

Note that after output of one row of the buffer has been performed, an "end of record" mark is placed on the external medium, whereupon that row of the buffer is filled with blanks (statement *blanks*).

50.3.2. Let us now assume that we have a code procedure *dataps* for transforming numerical values (according to a given format f) into decimal data in pseudostring notation:

«**procedure** *dataps* (x, f) *res*: (w) ;
    value x ;
    real x ; **integer array** w ; **string** f ;
    code».

Here x denotes the number to be transformed, w the resulting equivalent in pseudostring notation, and f indicates the format into which x should be brought. The meaning of f is exhibited by the following example: The string

$$'x = \sqcup + 099.999 \sqcup \sqcup \,'$$

means that e.g. the value $3.1415926535_{10}0$ should be output as

$$\boxed{x = \quad +03.142 \quad} \, .$$

50.3.3. In order to perform output of numerical data by means of *dataps*, the latter is incorporated into a procedure *insert* which allows inserting the converted data into the buffer (of course, in practice we would declare the combination *dataps/insert* directly as a code procedure):

«**procedure** *insert* (x, f, a) *trans*: (b) ;
    value a, x ;
    integer a, b ; **real** x ; **string** f ;
    comment uses global operand: **integer array** *buffer* ;
    begin
        integer k ;
        integer array $w\,[0: length\,(f)]$;
        procedure *dataps* (x, f) *res*: (w) ;
            value x ;
            real x ; **integer array** w ; **string** f ;
            code ;
        dataps (x, f) *res*: (w) ;
        for $k := 1$ **step** 1 **until** *length* (f) **do**
                buffer $[a, b + k - 1] := w\,[k]$;
        $b := b + length\,(f)$
    end *insert*».

Obviously this procedure has the effect that the decimal form of x is inserted into line a and columns $b, b + 1, \ldots$ of the buffer, whereupon

b is increased by the length of the inserted string such that it is just ready for the next output. However, it is up to the user to check whether the insertion is really inside the buffer and does not overwrite any other information.

50.3.4. As an example, let us draw a curve $y=f(x)$ with the convention that $y=p\times_{10}-2+q\times_{10}-3$ (p, q integer) be printed as the digit q in the p-th column of a line printer, while the argument x is recorded at the extreme left or right of every line, depending on where the curve runs:

```
« ; comment buffer is assumed to be filled with blanks and
                buffer [k, 0] = 120 for k = 1, 2, ..., 25 ;
  for x := 0, x while x < a do
  begin
      for k := 1 step 1 until 25 do
      begin
          y := 100 × f (x) ;
          p := entier (y) ;
          q := entier (10 × (y − p)) ;
          buffer [k, p] := q + 1 ;
          l := if p > 50 then 5 else 95 ;
          insert (x, '009.99999', k) trans : (l) ;
          x := x + h ;
      end k ;
      outbuf (15, 1, 25) ;
      comment it is assumed that channel number 15 corresponds to a
                line printer ;
  end x ».
```

50.3.5. Further example: computation and printing of the first 20 lines of the Pascal triangle (also given as an example at the end of the Knuth report [22]):

```
« ; comment buffer is assumed to be filled with blanks and
                buffer [k, 0] = 120 for k = 1, 2, ..., 25 ;
  begin
      integer j, k, l ;
      integer array p [−1 : 19] ;
      p [−1] := 0 ;
      for j := 0 step 1 until 19 do
      begin
          comment produce j + 1-th line of Pascal triangle ;
          l := 58 − 3 × j ;
          p [j] := 1 ;
```

for $k := j - 1$ **step** -1 **until** 0 **do** $p[k] := p[k] + p[k-1]$;
for $k := 0$ **step** 1 **until** j **do**
                    $insert(p[k], '\sqcup 00009', j+1)$ $trans:(l)$
end j ;
$outbuf(15, 1, 20)$;
comment it is assumed that channel number 15 is associated with
                    a 120-column line printer ;
end».

§ 51. Use of *inarray*, *outarray* for Auxiliary Storage

For solving numerical problems involving large arrays that cannot
be held in the high speed store, the standard procedures *inarray* and
outarray, together with the control operations mentioned in 49.4, prove
extremely useful. In all examples given here, we shall use the convention
of 49.4.2 (explicit resetting).

51.1. Choleski decomposition of a large matrix[1]

51.1.1. Let $a[1:n, 0:m]$ be an array representing a symmetric n-th order
bandmatrix (that is, one with $a_{ik}=0$ for $|i-k|>m$) in the usual band-
matrix representation (cf. 36.6). $a[i, k]$ therefore denotes the matrix
element in the i-th row and $i+k$-th column. It is assumed that m is
small compared to n, a representative example being $n=2000$, $m=50$.

The elements of the array $a[1:n, 0:m]$ are assumed to be recorded
row by row, i.e. in the order

$$a[1, 0], a[1, 1], \ldots, a[1, m], a[2, 0], \ldots, a[2, m], a[3, 0], \ldots, a[n, m],$$

as real numbers on a tape corresponding to channel number 56.

Note that the elements $a[i, k]$ with $i+k>n$, which do not correspond
to proper matrix elements, must be recorded as zeros.

Our task is to decompose the matrix A represented by the array a
into $A=R^T R$, where R is an upper triangular matrix represented as
an array $r[1:n, 0:m]$. Thus bandmatrix notation is used also for R, but
whereas for both matrices R and A the subdiagonal elements need not
be considered, this has different reasons; indeed, the subdiagonal elements of R vanish, while A is symmetric.

51.1.2. The present program is designed such that the array a is read
into the computer row by row, the elimination process executed for
every row of a, and the resulting rows of r again recorded on a tape
associated with channel number 57. To compute the elements of r, the
contributions that must be subtracted from $a[i, k]$ in the elimination

---

[1] Compare also section 6.3 in [*44*].

process are collected in the component $z[i, k]$ of an auxiliary array z. As a consequence, we have

$$r[i, 0] := sqrt(a[i, 0] - z[i, 0]) ;$$
$$r[i, k] := (a[i, k] - z[i, k])/r[i, 0] ; \qquad (k = 1, 2, 3, \ldots, m).$$

After computing these components, we add the products $r[i, p] \times r[i, q]$ to $z[i+p, q-p]$, and this for all $p = 0, 1, \ldots, m, q = p, p+1, \ldots, m$. In our program, however, we shall avoid the occurrence of both the arrays a and r and store at any one time only one row of a or r. For this purpose we declare an array $b[0:m]$.

Unfortunately, it appears now that in our program the array $z[1:n, 0:m]$ must be declared, which is just as big as a. This waste of storage space can be avoided, however, if we make use of the fact that by virtue of the bandform the elimination process involves at any moment at most $m + 1$ rows of the matrix a. Accordingly, it suffices to declare the array z as «**array** $z[1:m, 0:m]$», the rows of which are used cyclically. More precisely, what before was stored as $z[i, t]$ will now be stored as $z[s, t]$, where

$$s \equiv i \text{ (modulo } m), \quad \text{and} \quad 1 \leq s \leq m.$$

51.1.3. With this, the following program is obtained:
«**begin**
    integer i, j, k, l, p, t, s ;
    array $b[0:m], z[1:m, 0:m]$;
    $p := 1$;
    outsymbol $(56, \text{'rewind A-tape'}, -12)$;
    outsymbol $(57, \text{'rewind R-tape'}, -12)$;
    for $i := 1$ **step** 1 **until** m **do**
        for $j := 0$ **step** 1 **until** m **do** $z[i, j] := 0$;
grand loop:
    for $k := 1$ **step** 1 **until** n **do**
    begin

inarray $(56, b)$;	{ Read in k-th row of matrix A.

        for $j := 0$ **step** 1 **until** m **do**
            $b[j] := b[j] - z[p, j]$;
    if $b[0] \leq 0$ **then**
    begin

for $j := 1$ **step** 1 **until** 28 **do**
            outsymbol $(15, \text{'matrix} \sqcup not \sqcup positive$
                        $\sqcup definite\text{'}, j)$;
      goto *exit*
    end *if* ;
	Measures taken if
	matrix A is not posi-
	tive definite (this is
	signaled by a non-
	positive $b[0]$).

    $b[0] := sqrt(b[0])$;

for $j := 1$ **step** 1 **until** m **do**
$$b[j] := b[j]/b[0] ;$$

outarray $(57, b)$;
outsymbol $(57, 'end\ of\ record', -1)$;

> Output of k-th row of matrix R together with subsequent "end of record" mark.

for $j := 0$ **step** 1 **until** m **do** $z[p, j] := 0$;

> Clear p-th row of matrix Z (this would be the k-th row of the array $z[1:n, 0:m]$ in the noncyclic arrangement).

$l := 0$;
for $s := p+1$ **step** 1 **until** m,
                              1 **step** 1 **until** p **do**
begin
  $l := l+1$;
  for $t := 0$ **step** 1 **until** $m-l$ **do**
      $z[s, t] := z[s, t] + b[l] \times b[l+t]$;
end s ;
$p := p+1$;
if $p > m$ **then** $p := 1$;
end k ;
outsymbol $(57, 'rewind\ R\text{-}tape', -12)$;
exit:
 end *of program* ».

> This is the addition of $r[i, p] \times r[i, q]$ to $z[i+p, p-q]$, but transcribed into the cyclic arrangement.

51.1.4. After termination of this program, the matrix R is in band-matrix notation on tape 57, which is already rewound and therefore can be used immediately for solving the linear system

$$A \vec{x} = \vec{c}.$$

Under the assumption that the constant terms are already stored as a one-dimensional array $c[1:n]$, the following piece of program computes the solution x and prints it via channel 15:

«**begin**
    comment forwardsubstitution ;
    array $b[0:m]$;
    for $k := 1$ **step** 1 **until** n **do**
    begin
        inarray $(57, b)$;

> Read k-th row of matrix R.

        $c[k] := c[k]/b[0]$;
        $max := $ **if** $k+m > n$ **then** $n-k$ **else** m ;

for $j := 1$ **step** 1 **until** *max* **do**
$$c[k+j] := c[k+j] - b[j] \times c[k] ;$$
end k ;
comment backsubstitution ;
for $k := n$ **step** -1 **until** 1 **do**
begin
  outsymbol $(57,$ '*rewind R-tape to begin of*
                     row k', $-11)$;

  inarray $(57, b)$; $\begin{cases} \text{Read } k\text{-th row of} \\ \text{matrix } R. \end{cases}$

  $max := $ **if** $k+m>n$ **then** $n-k$ **else** m ;
  for $j := 1$ **step** 1 **until** *max* **do**
    $$c[k] := c[k] - c[k+j] \times b[j] ;$$
  $c[k] := c[k]/b[0]$;
  outsymbol $(57,$ '*rewind R-tape to begin of row k*', $-11)$;
end k ;
outsymbol $(15,$ '*new page*', $-2)$;
for $j := 1$ **step** 1 **until** $8, -1$ **do** *outsymbol* $(15,$ '*solution*', $j)$;
outarray $(15, c)$;
end».

51.2. High order qd-algorithm

51.2.1. One step of the *qd-algorithm*, more precisely of its *progressive form* which allows the computation of eigenvalues (cf. [*31*]), can be described by the following statements:

«$q[1] := q[1] + e[1]$;
for $k := 1$ **step** 1 **until** $n-1$ **do**
begin
  $e[k] := (e[k]/q[k]) \times q[k+1]$;
  $q[k+1] := (q[k+1] - e[k]) + e[k+1]$
end».

Here n denotes the order of the *qd*-table and $e[n]$ is assumed to be zero. Note that the q- and e-values of the new line have the same names as those of the original *qd*-line; the original values are therefore destroyed.

If the above piece of program is executed iteratively, then the e's converge to zero and the $q[k]$ to the desired eigenvalues provided certain conditions are fulfilled (for instance, if at the beginning all q's and e's are positive). It is this fact which makes the *qd*-algorithm feasible for computing eigenvalues of tridiagonal matrices and poles of continued fractions.

51.2.2. However, in certain applications the order n is so high that the use of auxiliary storage must be envisaged. To this end it is assumed that initially the given q- and e-values are recorded interlaced on a magnetic tape:

$$q[1], e[1], q[2], e[2], \ldots, e[n-1], q[n], e[n] = 0. \qquad (1)^{1}$$

Furthermore, we require that all q- and e-values be positive because otherwise numerical instability and other troubles might occur. To perform one step of the qd-algorithm we take the elements of the interlaced array (1) into the computer in groups of 1000, apply the qd-formulae to these and put the new q- and e-values back onto another tape, which after a full step contains the elements of the new qd-line. In the next step we operate from the second tape back to the first, etc.

The formulae used are essentially the same as in 51.2.1, except that the 1000 elements read in are stored as one array $qe[1:999]$ plus one simple variable e, both together containing 500 q- and e-values. The extra variable e is needed to make the connection with the next group of 1000 elements. In order to exhibit the details, we picture the operations to be performed by means of the rhombus rules[2]:

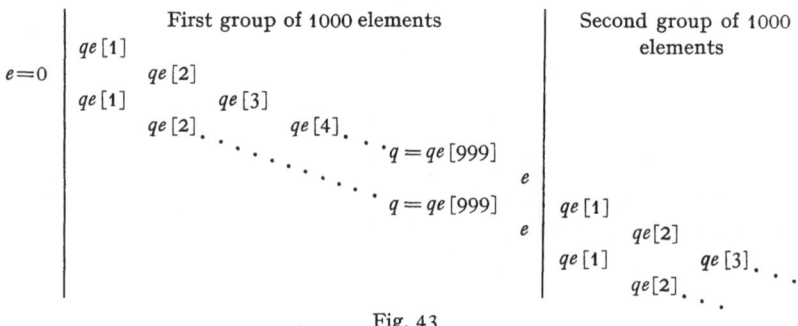

Fig. 43

As is obvious, the transition from group to group requires special care insofar as the new $qe[999]$ cannot be computed before e has been read in, while at the same time the new e depends upon $qe[1]$ of the next group. Accordingly, in- and output of the components of qe and of the variable e must occur in appropriate order.

The following procedure $qdtape$ executes one full qd-step, that is, it computes for given q- and e-values read from a tape a the q- and e-values for the next qd-line and records them on a tape b. On both tapes the ordering is supposed to be the same as in (1).

```
«procedure qdtape (n, a, b) ;
    value n, a, b ;
    integer n, a, b ;
```

---

[1] $e[n]$ must always be zero.
[2] For the definition of the rhombus rules see § 5 in [35].

```
      begin
        integer k, j, p ;
        real q, e ;
        outsymbol (a, 'rewind tape a', − 12) ;
        outsymbol (b, 'rewind tape b', − 12) ;
        for k := 0 step 1000 until 2×n−3 do
loop:   begin
```

> $p =$ number of elements in the next group to be read. For every group, p is chosen such that $p = 1000$, except for the last group, which may contain between 4 and 1002 elements. 1002 is chosen as upper limit in order to avoid the last group containing only 2 elements since this would cause trouble.

$$p := \text{if } 2 \times n - k \leqq 1002$$
$$\text{then } 2 \times n - k \text{ else } 1000 \ ;$$

```
          begin
            array qe[1 : p−1] ;
            inarray (a, qe) ;
```

> Read bulk of group.

$$e := \text{if } k = 0 \text{ then } 0 \text{ else } (e/q) \times qe[1] \ ;$$
$$\text{if } k \neq 0 \text{ then } outreal(b, e) \ ;$$

> Rhombus rule for last element e of previous group and output of the new e.

$$qe[1] := (qe[1] - e) + qe[2] \ ;$$
```
            for j := 2 step 2 until p−4 do
            begin
```
$$qe[j] := (qe[j]/qe[j-1])$$
$$\times qe[j+1] \ ;$$
$$qe[j+1] := (qe[j+1] - qe[j])$$
$$+ qe[j+2]$$
```
            end j ;
```
$$qe[p-2] := (qe[p-2]/qe[p-3])$$
$$\times qe[p-1] \ ;$$

> Rhombus rules for bulk of group.

$$inreal(a, e) \ ;$$
$$q := qe[p-1] :=$$
$$(qe[p-1] - qe[p-2]) + e \ ;$$

> Take in the last element e of group, apply rhombus rule to the last q-element $(qe[p-2])$ of the group.

$$outarray(b, qe)$$

> Output of bulk of group.

```
          end block
        end k ;
        outreal (b, 0) ;
```

> Output of last element e of group.

```
      end qdtape».
```

16*

To use this procedure, it is called repetitiously, whereby the user must interchange the values of a and b for every call.

51.3. Matrix inversion by the escalator method

51.3.1. Let A be an infinite symmetric matrix, all finite principal submatrices of which are positive definite. We want to invert a finite section A_n consisting of the first n rows and columns of A. If n is not known a priori but depends on the behavior of A_n^{-1} as $n \to \infty$, then by virtue of rule 39.3.3 a difficulty arises insofar as n is unknown at the moment the array $a[1:n, 1:n]$ carrying the elements of A_n should be declared.

In *full* ALGOL there is a device which allows overcoming this difficulty, namely the *own-feature*[1]. Unfortunately, however, most implementors of ALGOL exclude just that part of this feature which could help in this problem, namely the so-called *dynamic own-arrays*.

We can, however, solve our problem by means of the standard I/O-procedures. To this end we choose the *escalator method* for inverting A_n since this method allows to compute all inverses $A_1^{-1}, A_2^{-1}, A_3^{-1}, \ldots, A_n^{-1}$ one after the other with an effort that is proportional to n^3. We can therefore continue the calculation just as long as it is necessary to achieve the desired effect.

51.3.2. The numerical process can be described as follows: Let

$$A_k = \left\{ \frac{A_{k-1} \mid \vec{b}}{\vec{b}^T \mid c} \right\}, \qquad A_k^{-1} = \left\{ \frac{X \mid \vec{y}}{\vec{y}^T \mid z} \right\},$$

while $A_{k-1}^{-1} = F$ is already known. Here

$$
\begin{aligned}
F, X, A_{k-1} \quad & \text{are } (k-1) \times (k-1)\text{-matrices} \\
\vec{w}, \vec{h}, \vec{y} \quad & \text{are } (k-1)\text{-vectors} \\
s, c, z \quad & \text{are scalars.}
\end{aligned}
$$

Then with $\vec{w} = F\vec{b}$, $s = \vec{b}^T \vec{w}$, we have

$$
\begin{aligned}
z &= 1/(c-s), \\
X &= F + z \times \vec{w} \times \vec{w}^T, \\
\vec{y} &= -z \times \vec{w}.
\end{aligned}
$$

In designing the program, we must be aware that k can run to such high values that storing a $k \times k$-matrix becomes impossible; therefore we must keep all matrices on tape, reading only row- or column-vectors into the high speed storage.

---

[1] Since the own-feature is not treated in this Handbook, the reader is referred to other books on the subject, e.g.: [53].

In contrast to earlier examples, we now make a more sophisticated use of the channel number: Indeed, it is assumed that at the beginning of step k the elements of $F = A_{k-1}^{-1}$ are on external media such that (for every $l = 1, 2, \ldots, k-1$) the elements

$$f_{l1}, f_{l2}, f_{l3}, \ldots, f_{ll}$$

are ready for input through channel $100 + l$. Note that F is symmetric so that the superdiagonal elements need not be given.

Now we proceed as follows: The elements $a_{k1}, a_{k2}, \ldots, a_{kk}$ of the k-th line of the matrix A are generated and assigned to the array $b[1:k]$ by a generator procedure *gener*. Then the elements of F are read in row by row, and at the same time the product $\vec{w} = F\vec{b}$ is computed and stored as an array $y[1:k-1]$; after that, z and s are computed. Then the elements of F are again read in row by row but this time for computing X, the rows of which are again recorded row-wise, namely the l-th row via channel $100 + l$. Finally the vector $\vec{y}$, which is the k-th row of A_k^{-1}, is computed and recorded via channel $100 + k$. This completes step k.

51.3.3. Our program is designed as a procedure *escal*, whose formal operands n, *gener*, *stop*, have the following meaning:

«**integer** n» indicates at the beginning the maximum value which k should attain and after termination of the process the last value of k; it is therefore the order of the last inverse A_k^{-1}.

«**procedure** *gener* (n, x) ; **value** n ; **integer** n ; **array** x ; **code**»

is a procedure with the property that «*gener* (k, b)» generates the values $b[1], b[2], \ldots, b[k]$, which are the subdiagonal and diagonal elements of the k-th row of A $(b[i]$ is in fact $a[k, i])$.

«**Boolean procedure** *stop* (k, l, al) ;
        value k, l ; **integer** k, l ; **array** al ; **code**»

serves to make a decision for terminating the process. The values $al[1], al[2], \ldots, al[l]$, which are the diagonal and subdiagonal elements of the l-th row of A_k^{-1}, can be used to make the decision.

With these hypotheses the following program emerges:

«**procedure** *escal* (*gener*, *stop*) trans: (n) ;
    integer n ; **procedure** *gener* ; **Boolean procedure** *stop* ;
    begin
        real s, z ;
        integer k, l, j ;
        for $k := 1$ **step** 1 **until** n **do**

```
begin
   array b, y[1:k] ;
   gener (k, b) ;
   outsymbol (100 + k, 'rewind row k', −12) ;
   for l := 1 step 1 until k do y[l] := 0 ;
   for l := 1 step 1 until k−1 do
   begin
      array al[1:l] ;
      inarray (100 + l, al) ;
      if stop (k−1, l, al) then goto finis ;
      for j := 1 step 1 until l−1 do
         y[j] := y[j] + al[j] × b[l] ;
      for j := 1 step 1 until l do
         y[l] := y[l] + al[j] × b[j] ;
      outsymbol (100 + l, 'rewind row l',
                                         −12) ;
   end l ;
   s := 0 ;
   for j := 1 step 1 until k−1 do
         s := s + b[j] × y[j] ;
   z := y[k] := 1/(b[k] − s) ;
   for l := 1 step 1 until k−1 do
   begin
      real z1 ;
      array al[1:l] ;
      inarray (100 + l, al) ;
      z1 := z × y[l] ;
      for j := 1 step 1 until l do
         al[j] := al[j] + z1 × y[j] ;
      outsymbol (100 + l, 'rewind row l',
                                         − 12) ;
      outarray (100 + l, al) ;
      outsymbol (100 + l, 'rewind row l', −12) ;
   end l ;
   for j := 1 step 1 until k−1 do
            y[j] := − z × y[j];
   outarray (100 + k, y) ;
   outsymbol (100 + k, 'rewind row k', −12) ;
end k ;
goto out ;
```

Annotations (right margin):

- `gener (k, b)` : Generate k-th row of matrix A.
- `outsymbol (100 + k, ...)` : Reset channel for recording k-th row of A_k^{-1}.
- Braced group (inarray/for loops): Compute $\vec{y} = F\vec{b}$ taking into account the special arrangement of matrix F on the external medium.
- `s`, `z` group: Compute z.
- `inarray (100 + l, al)` group: Read l-th row of matrix F.
- Large group: Compute l-th row of matrix X, record this row via channel $100 + l$ (thus overwriting l-th row of matrix F) and reset this channel again.
- `outarray (100 + k, y)` group: Record k-th row of matrix A_k^{-1}.

finis: $n := k-1$;
out:
    end *escal*».

Note that after termination, all channels 1 through n, which contain the rows 1 through n of the inverse of the matrix A_n, are reset.

51.3.4. How should this procedure now be used?

As an example, let the infinite matrix A be $H+zI$, where H is the Hilbert matrix and I the infinite unit matrix, while z is a real-valued parameter. Then if the process should be continued as long as the 1, 1-element of A_k^{-1} (as a function of k) still changes its value, the actual counterparts of *gener* and *stop* must be declared as follows:

```
«procedure gener1 (n, b) ;
    value n ;
    integer n ; array b ;
    begin
        integer l ;
        for l := 1 step 1 until n−1 do b[l] := 1/(l+n−1) ;
        b[n] := z+1/(2×n−1)
    end gener».
«Boolean procedure stop1 (k, l, al) ;
    value k, l ;
    integer k, l ; array al ;
    if k=0 then
            begin x11 := 0 ; stop1 := false end
    else
        if l=1 ∧ al[1] ≠ x11 then
            begin x11 := al[1] ; stop1 := false end
        else
            if l=1 then stop1 := true
            else
                stop1 := false».
```

Appendix A

§ 52. The Jensen Device

In 37.5 and 47.4.2 difficulties were encountered when a procedure, originally designed for operating upon one-dimensional arrays, could not be used to perform the same actions upon rows or columns of two-dimensional arrays. As mentioned already, this difficulty can be overcome by means of the so-called *Jensen device*[1]. Since this feature, though not possible in the SUBSET, is sometimes really useful, it will be described here.

52.1. The full name-concept

The rather stringent rule of 45.3.2 concerning actual parameters has a far less restrictive analogue in full ALGOL; indeed, if a formal parameter is called by name and specified as a simple variable, then in full ALGOL the corresponding actual parameter may (in contrast to the SUBSET) be an expression of the appropriate type. This has the consequence that by virtue of the substitution rule of 45.2.2 (which holds also in full ALGOL) this expression appears in the equivalence block wherever the formal parameter occurred at a corresponding place in the procedure body, e.g.[2]:

Declaration: «**procedure** $x(a, b)$; **real** a, b ; $a := b$».
Call: «$x(c[k], p + 1)$».
Equivalence block: «**begin real** $æ$; $c[k] := (p + 1)$ **end**».

Alternatively, procedure *nevint* may be taken as declared in 44.7.4 and called by

«*nevint* (n, aa, bb, z) *res*: $(c[k])$».

This call has the effect that the interpolated value is assigned directly to the subscripted variable. As we know, this is not possible in the SUBSET since there a subscripted variable cannot be used as actual counterpart of a formal parameter called by name. In the SUBSET therefore, the above operation must be done in two steps:

«*nevint* (n, aa, bb, z) *res*: (aux) ;
$c[k] := aux$».

---

[1] Proposed in 1960 by J. Jensen of the Regnecentralen, Copenhagen.

[2] Where the formal parameter occurs as a primary in an expression, the corresponding actual parameter must of course, after insertion into the equivalence block, also have the status of a primary. To this end, wherever needed, the actual parameter must be enclosed in parentheses before performing the substitution.

Of course, one must be very careful in using the full name-concept, since even in full Algol the equivalence block must in any case be a meaningful piece of program. Thus

$$\text{«}nevint\,(n,\, aa,\, bb,\, z)\; res:(a+b)\text{»}$$

is not allowed since the last statement in the equivalence block would then become «$a+b := y[n]$», which is complete nonsense.

It would seem that a general rule to avoid such occurrences is that a general expression may appear as actual parameter in a procedure call only if the corresponding formal operand is an argument of the procedure. However, this is no longer possible in full Algol since there the classification of operands as arguments, transients, results and exits is not appropriate; in fact, we cannot even speak of operands of a procedure. This means that in full Algol the only criterion for the legality of a procedure call is that the equivalence block must become a meaningful piece of program.

52.2. The Jensen device

52.2.1. Let us declare a function procedure as follows:

```
«real procedure sum (a, b, c, d) ;
    real c ; integer a, b, d ;
    begin
        real s ;
        s := 0 ;
        for d := a step 1 until b do s := s+c ;
        sum := s
    end sum».
```

On first sight this does not make much sense; indeed, as long as we adhere to the Subset rule admitting only identifiers as actual counterparts of formal parameters called by name, a function designator «$sum\,(w,\, x,\, y,\, z)$» can only produce the value $(x-w+1)\times y$ (or zero).

However, in full Algol sum may be called e.g. by

$$\text{«}sum\,(1,\, n,\, 1/a[k],\, k)\text{»}, \tag{1}$$

for which the substitution rule immediately yields the equivalence block

```
«begin
    real s ;
    s := 0 ;
    for k := 1 step 1 until n do s := s + (1/a[k]) ;
    sum := s                                  substituted for c
end».
```

Accordingly, the call (1) computes $\displaystyle\sum_{k=1}^{n} \frac{1}{a_k}$.

The essential point (and the content of the Jensen device) is that the third actual parameter $1/a[k]$ formally depends upon the fourth actual parameter k; however, since the latter undergoes changes during the execution of the procedure, $1/a[k]$ also takes part in these changes and thus a nontrivial effect is achieved.

52.2.2. In a similar way, an inner product procedure may be declared as

```
«procedure innerp (n, x, y, s, k) ;
    value n ;
    real s, x, y ; integer n, k ;
    begin
        s := 0 ;
        for k := 1 step 1 until n do s := s + x × y ;
    end innerp».
```

Again this looks rather queer but makes sense as soon as one considers

$$\text{«}innerp\,(k, p[j], q[j], x, j)\text{»}. \tag{2}$$

Indeed, the corresponding equivalence block is

```
«begin
    real æ ;
    x := 0 ;
    for j := 1 step 1 until k do x := x + p[j] × q[j]
    end equivalence block»,
```

and therefore the call (2) does exactly the same thing as the call «$inner\,(k, p, q, x)$» given in 44.1.3. However, $innerp$ can be called just as well by

$$\text{«}innerp\,(l, a[i, j], a[j, k], t, j)\text{»}, \tag{3}$$

for which the substitution rule yields the equivalence block

```
«begin
    real æ ;
    t := 0 ;
    for j := 1 step 1 until l do t := t + a[i, j] × a[j, k]
    end».
```
                                        substituted   substituted
                                          for x   for y

Thus, obviously, the call (3) achieves what was unable to be accomplished by means of procedure $inner$ as declared in 44.1.3, namely computing $\sum_{j=1}^{l} a[i, j] \times a[j, k]$ directly by a call of $inner$. (3) again exhibits the mechanism of JENSEN's idea, namely that (as an example)

the second and third actual parameter formally depend upon the fifth actual parameter.

52.2.3. In a similar way, the full name-concept can be applied to examples involving arbitrary functions: If it is desired to integrate a differential equation $y' = f(x, y)$ by means of the Euler method (cf. 48.1.3), one can declare a procedure

```
«procedure jeuler (x, h, p, f, y, yy) ;
    value h, p ;
    real h, x, f, y ; integer p ; array yy ;
    begin
        integer j ;
        for j := 1 step 1 until p do
        begin
            yy[j] := y := y + h×f ;
            x := x + h
        end j ;
    end jeuler».
```

This procedure serves to integrate a differential equation $y' = f(x, y)$ with given initial values x, y, in p steps of length h, yielding the solution in tabular form as an array $yy[1:p]$. However, here the function f is not to be defined as a function procedure but as an arithmetic expression to be given as actual counterpart of the formal parameter f. As an example,

$$«jeuler (t, 0.01, 100, t{\uparrow}2 + p{\uparrow}2, p, q)»$$

will integrate $\dfrac{dp}{dt} = t^2 + p^2$ in 100 steps of length 0.01, the current values of t, p providing the initial values, while the solution is stored as an array $q[1:100]$.

52.2.4. Thus this feature, called the *Jensen device*, is certainly an elegant instrument, but it should not be overlooked that — aside from questions of economy — it can be applied only in connection with procedures that have been designed especially with this application in mind. Indeed, let e.g. pp be a procedure with an array a as formal operand:

```
«procedure pp (n, a) ;
    value n ;
    integer n ; real array a ;
    begin
        integer i, j ;
        for i := 1 step 1 until n do
            for j := 1 step 1 until n do
                a[i−j] := a[i] × a[j]
    end pp».
```

For adapting this procedure to the Jensen device in order to allow that a subarray of an array of higher dimension than *a* may be used as actual counterpart of *a*, one must

a) Not only replace every occurrence of a subscripted variable $a[\ldots]$ by a simple variable *a*, but use different variables *a1, a2, a3,* ... for syntactically different subscripted variables corresponding to the array *a*.

b) List all these variables *a1, a2,* ... as formal parameters of the procedure.

c) List all internal variables of *pp*, as far as they occur in subscripted variables $a[\ldots]$, as formal parameters of the procedure (this is required because of the rule given in 45.2.3).

For the above procedure *pp* this would mean that it must be re-written as

> «**procedure** $jj\,(n,\,a1,\,a2,\,a3,\,i,\,j)$;
>   **value** n ;
>   **real** $a1,\,a2,\,a3$; **integer** $n,\,i,\,j$;
>   **begin**
>   **for** $i := 1$ **step** 1 **until** n **do**
>   **for** $j := 1$ **step** 1 **until** n **do**
> $a3 := a1 \times a2$
>   **end** jj ».

Now, if procedure jj is to be called in such a way that the same effect as by a call «$pp\,(n,\,a)$» is achieved, one must write

$$«jj\,(n,\,a[i],\,a[j],\,a[i-j],\,i,\,j)»,$$

but the same action can just as well be performed upon the *l*-th component of a matrix *b*, namely by a call

$$«jj\,(n,\,b[l,\,i],\,b[l,\,j],\,b[l,\,i-j],\,i,\,j)».$$

This follows immediately from the substitution rule which in the latter case yields the equivalence block

> «**begin**
>   **integer** $næ$;
> $næ := n$;
>   **begin**
>   **for** $i := 1$ **step** 1 **until** $næ$ **do**
>   **for** $j := 1$ **step** 1 **until** $næ$ **do**
> $b[l,\,i-j] := b[l,\,i] \times b[l,\,j]$
>   **end**
> **end** *equivalence block*».

It should be recognized, however, that the flexibility of the Jensen device allows a nearly unlimited number of other applications of procedure jj; as an example, the matrix a could be computed as the dyadic product of the vectors b and c by a call

$$« jj\,(n,\,b\,[p],\,c\,[q],\,a\,[p,\,q],\,p,\,q)\,».$$

52.2.5. *Disadvantages.* On the other hand, the price to be paid for this extra flexibility is that the call of a procedure involving the Jensen device is complicated even if the extra flexibility is not needed. In fact, such a procedure statement can be written only with careful consideration of the procedure body, a necessity which not only contradicts the intention of the procedure concept altogether, but may truly be a burden for a procedure with a very extended body. As a consequence it is recommended to use the Jensen device only in situations where it is really indispensable.

52.3. Bound variables

In view of these disadvantages it is probably worthwhile to investigate the background of the Jensen-device a little closer in order to assign it its proper place within the framework of ALGOL:

52.3.1. Considering a mathematical expression, e.g. an integral

$$I = \int_0^1 f(x, y)\, dx,$$

it is seen that the variables x and y serve entirely different purposes: y is a variable upon which the value of I depends; in mathematical logic this is called a *free variable* (of the expression). This latter term indicates that one is free to substitute a value, e.g. 2.75, for y, whereupon one obtains the result

$$\int_0^1 f(x, 2.75)\, dx.$$

The variable x, on the other hand, is only an auxiliary object for describing the operation to be performed by the expression. It is called a *bound variable* since it is not accessible from outside the expression. Indeed, it would be stupid to substitute in the above integral the value 2.75 for x.

Likewise in the expression

$$\sum_{k=1}^l a_k\, b_k$$

k is a bound variable while l, a, b are free.

52.3.2. Comparing these examples with an ALGOL procedure and the terminology used in Chapter VII, it becomes obvious that the free variables correspond to what are called the operands of a procedure, while the bound variables correspond to the internal quantities.

With these notions in mind, a glance at the declarations of the procedures *sum, innerp, jeuler, jj* reveals at once the essence of the Jensen device: *Bound variables of a procedure are quoted in the formal parameter part* in order that actual parameters may be made dependent upon them.

52.3.3. However, since the occurrence of bound variables in the formal or actual parameter part is an irregularity (but cannot be avoided in this context) it is recommended that such occurrences at least be clearly indicated, namely by introducing a fifth category in the structurized formal and actual parameter part, e.g.

«**real procedure** *sum* (a, b, c) *bound variables*: (d) ; ...»,
«**procedure** *innerp* (n, x, y) *res*: (s) *bound variables*: (k) ; ...»,
«**procedure** *jj* $(n, a1, a2)$ *res*: $(a3)$ *bound variables*: (i, j) ; ...»,
«*sum* $(1, n, 1/a[k])$ *bound variables*: (k)»,
«*innerp* $(n, a[i, j], b[j, k])$ *res*: $(c[i, k])$ *bound variables*: (j)».

§ 53. Conclusion

In view of its ad-hoc character it seems doubtful that the Jensen device (and to some extent even the full name-concept) is the last word in programming language design. Indeed, the dependence of the components of an array upon its subscripts (and likewise the dependence of a function upon its arguments) is more appropriately described by means of CHURCH's lambda notation[1] rather than through the bound variables of a computing process. Accordingly, we conclude with a sideview to a possibility for introducing this notation in a future ALGOL, but in doing so we strictly adhere to a SUBSET like language-concept, i.e. one in which quantities rather than names play the fundamental role.

53.1. Church's lambda notation[1]

Let us consider again

$$I = \int_0^1 f(x, y)\, dx,$$

---

[1] CHURCH, A.: A Set of Postulates for the Foundations of Logic. Ann. of Math. II **33**, 346—366 (1932).

The colons following the lambda-variables have been added by the present author.

which represents a certain value depending on the value of y. In standard mathematical notation the same form of expression is used for describing the value of this integral for a given y as well as for the functional relationship (as an abstract entity) between y and I. In view of confusions that may arise from this ambiguity, A. CHURCH proposed different notations for the two concepts:

a) by $\int_0^1 f(x, y)\, dx$ he describes the value of the integral for a given value of y, while

b) $\lambda y: \int_0^1 f(x, y)\, dx$ denotes the function $I(y)$ itself, i.e. the functional relationship between the set of all admissible y's and the corresponding I's[1].

In this manner $\lambda x: sin(x)$ denotes the *sine-function* itself while $sin(x)$ represents the *value* of the sine-function for a specific value of x.

Finally also $\lambda x: (a_0 + a_1 x + a_2 x^2 + \cdots + a_n x^n)$ is the correct expression of a polynomial, indicating that it is considered as a function of x, whereas if it should be considered as a linear form of its coefficients, it would have been written

$$\lambda a_0 a_1 a_2 \ldots a_n : (a_0 + a_1 x + a_2 x^2 + \cdots + a_n x^n).$$

53.2. The lambda notation for arrays

Now, since the components of an array are also functions of its subscripts, one should correctly write $\lambda i, k: a[i, k]$ for this function, i.e. for the array where it appears as a quantity. On the other hand

$a[i, k]$	would denote one single component of this array,
$\lambda i: a[i, j]$	would mean the j-th column vector,
$\lambda i, j: a[i+1, j+1]$	the shifted array,
$\lambda i: a[i, i]$	the vector of all diagonal elements, etc.

Obviously this notation — if it is used for the array components and not just for their values — provides the mechanism necessary for performing operations upon subarrays, and it can easily be transcribed into ALGOL:

As a first measure, a formal array must be represented in the formal parameter part of a procedure declaration not just by its identifier F

---

[1] This is an analogous distinction as in ALGOL, where on the one hand the declaration of a function procedure, e.g.

«**real procedure** $I(y)$; **real** y ; ...»

defines the functional relationship between y and I, while the value of the function for a given value of y is described by a function designator, e.g. «$I(y)$».

but by the syntactic construction

$$\text{«}\mathbf{lambda}\ V_1, V_2, \ldots, V_p\colon F[V_1, V_2, \ldots, V_p]\text{»},\tag{1}$$

where «**lambda**» is a new basic symbol, F is the array identifier, p the dimension of the array F and the V's are identifiers denoting hypothetical simple variables of type **integer**. These latter are assumed to be bound to the construction (1); as a consequence they need not be declared, not even specified.

A corresponding actual operand would then be described by

$$\text{«}\mathbf{lambda}\ V_1, V_2, \ldots, V_p\colon A[S_1, S_2, \ldots, S_q]\text{»},\tag{2}$$

where now the S's are subscript expressions while the V's are again identifiers representing integer-type simple variables which are bound to (2). q is the dimension of the array A, of which (2) is a subarray. The S's may depend on the V's (but also on other variables) and in this way they define the subarray[1].

For practical reasons a few exceptions must be allowed for:

a) In most applications an actual array occurring in a procedure call is an array with precisely the same dimension and with the same meaning of the subscripts as for the corresponding formal array. It would be very annoying if even in such trivial cases the actual operand had to be quoted in full, i.e. by

$$\text{«}\mathbf{lambda}\ V_1, V_2, \ldots, V_p\colon A[V_1, V_2, \ldots, V_p]\text{»}.$$

Indeed, in this case one could obviously just as well write «A» without losing anything.

b) In view of the standard I/O-procedures *inarray*, *outarray*, or procedures like *vgs* (given as an example in 45.4.4), it must be allowed that the dimension of a formal array is not specified by the procedure declaration. Therefore it must be tolerated that as in ALGOL 60 a formal array is quoted in the procedure heading just by its identifier F. However, if this is done, the corresponding actual parameter must also be an identifier, which makes it impossible to use a subarray as actual operand.

53.3. Syntax of the proposed extension

It is not actually necessary to introduce the full syntactic forms (1) and (2) above. Indeed, since in (1) neither the new symbol «**lambda**» nor the variables V_k in front of the colon convey any information, these

---

[1] The term *subarray* should not be taken too literally; indeed, as indicated in 53.5, it can also mean a shifted array, or even an array of higher dimension than the array from which it is extracted (see also example 53.5.4).

can be omitted and the formal parameter defined as having one of the following syntactic forms:

«*F*» (array identifier),
«*F*[**1, 2,** ..., **p**]» *(formal array designator)*,[1]

where the underlined integers **1, 2,** ... (called *lambda designators*) are basic symbols of the language which (in this order) represent the bound variables $V_1, V_2, ..., V_p$.

Likewise an actual parameter can have — besides the cases already mentioned in 26.2 — the following syntactic form:

«*A*[$S_1, S_2, ..., S_q$]» *(actual array designator)*,

the *S*'s now being subscript expressions which may contain as primaries (among other variables valid in the environment of the call) the lambda designators **1, 2,** ..., **p**, where p is the dimension of the corresponding formal array *F*.

Actual array designators may also appear as actual parameters where the corresponding formal parameter is a simple variable, but then the subscripts of the actual array designator may not contain any lambda designators.

53.4. Semantics of the proposed extension

Whenever an actual operand is represented by an identifier, the same rules as in 45.2 apply, but if an actual array designator «*A*[$S_1, S_2, ..., S_q$]» occurs as actual parameter, this has the following consequences:

a) With the exception of lambda designators, all primaries occurring within the *S*'s are evaluated at the time when this actual parameter is encountered, and the values thus obtained are substituted in place of these primaries.

b) The equivalence block corresponding to that call is constructed as stated in 45.2, except that within the procedure body any occurrence of an element «*F*[$E_1, E_2, ..., E_p$]» referring to the formal array *F* is replaced by «*A*[$S_1, S_2, ..., S_q$]», whereby every lambda designator **k** occurring within that actual array designator is replaced by the expression E_k. Needless to say, name conflicts must be treated in analogy to the rules given in 45.2.3.

Thus the present proposal bases essentially on the fact that all objects occurring within the actual parameter part in brackets are "taken per value", i.e. evaluated when the procedure call is encountered and then are kept constant throughout the execution of the procedure. As a consequence, a procedure always operates on clearly defined quantities, a

---

[1] The grotesque type letter **p** is to be considered as representant of the number p written in grotesque type.

fact which makes it possible to build compilers that produce more efficient object programs.

The present concept could of course be extended from arrays to procedures, thus making it possible for a sub-procedure of a procedure to be used as actual operand. However, this is not so urgent since the global parameters give us this possibility already in the present ALGOL.

53.5. Applications of the proposed extension

The following examples serve to show some of the possibilities of this extension, which, though resembling somewhat the empty subscript positions of ALGOL 58, goes far beyond ALGOL 58 in scope and versatility.

53.5.1. The declaration of procedure *inner* as given in 44.1.3 could now be rewritten as

«**procedure** *inner* $(n, x[1], y[1])$ *res*: (s) ;
    value n ;
    real s ; **integer** n ; **array** x, y ;
    begin
              procedure body as in 44.1.1.
    end».

With this modification

«*inner* $(l, a[i, 1], a[1, k])$ *res*: (t) »

has the same effect as was achieved with the Jensen device by the call (3) in § 52. Furthermore,

«**for** $i := 1$ **step** 1 **until** n **do** *inner* $(n, a[i, 1], x)$ *res*: $(y[i])$ »

describes the multiplication of a matrix $A = a[1:n, 1:n]$ with a vector $\vec{x} = x[1:n]$, yielding the product vector $\vec{y} = y[1:n]$.

53.5.2. Assume that

«**procedure** *gauss* (n, r) *trans*: $(a[1, 2], x[1, 2])$ *exit*: $(sing)$;
    value n, r ;
    integer n, r ; **array** a, x ; **label** *sing* ;
    code»

declares a procedure which solves simultaneously r linear systems with a common coefficient-matrix $A = a[1:n, 1:n]$, the r sets of constant terms being given at the beginning as the r columns of a matrix $X = x[1:n, 1:r]$, and that after termination the r solutions are available as the r columns of the same matrix X.

In ALGOL 60 the actual counterpart of x had to be declared as a two-dimensional array, also in the case $r = 1$, but with the present proposal we can call *gauss* by

$$«gauss\,(n,\,1)\;trans\!:(a,\,b\,[\mathbf{1}])\;exit\!:(fail)\,»,$$

where b is declared as «**array** $b\,[1\!:\!n]$». According to the extended sub-stitution rule, any occurrence of «$x\,[i,\,j]$» within the procedure body of *gauss* is replaced by «$b\,[i]$» and thus — in view of $r=1$ — the above call solves the one linear system as defined by the coefficient-matrix A and the constant terms $b\,[1],\,\ldots,\,b\,[n]$.

53.5.3. Consider

«**procedure** *matinv* (n) *trans*$:(a\,[\mathbf{1},\mathbf{2}])$ *exit*$:(fail)$;
    value n ;
    integer n ; **array** a ; **label** *fail* ;
    begin ⎫
             ⎬ procedure body as in 44.7.5.
    end». ⎭

If one wants to invert the capsized matrix

$$\left\{\begin{matrix} a_{nn} & \cdots & a_{n1} \\ \vdots & & \vdots \\ a_{1n} & \cdots & a_{11} \end{matrix}\right\}$$

(which may be desirable for numerical reasons), this would normally require that its components be assigned to the components of another matrix which is then inverted. However, with the present scheme, this can be achieved directly by a call

$$«matinv\,(n)\;trans\!:(a\,[n+1-\mathbf{1},\,n+1-\mathbf{2}])\;exit\!:(fail)\,»,$$

because in this case *matinv* operates on the component $a\,[n+1-i,\,n+1-j]$ instead of $a\,[i,\,j]$.

53.5.4. Suppose that a procedure *jacobi* for computing eigenvalues of a symmetric matrix has been declared as

«**procedure** *jacobi* (n) *trans*$:(a\,[\mathbf{1},\mathbf{2}])$;
    value n ;
    integer n ; **array** a ;
    code *performs diagonalisation of matrix* $a\,[1\!:\!n,\,1\!:\!n]$».

Since a procedure for the Jacobi method is usually designed such that it does not use or change the subdiagonal elements, it is only natural to think of rewriting the procedure in such a way that the storage space for the subdiagonal elements is actually saved. With the present proposal this saving can be achieved without rewriting the procedure, simply by calling it by

$$«jacobi\,(n)\;trans\!:(d\,[(\mathbf{1}-1)\times(n-\mathbf{1}/2)+\mathbf{2}])\,»,$$

where the array d is declared as «**array** $d[1:n \times (n+1)/2]$». Indeed, $k = (i-1) \times (n-i/2) + j$ is the function which maps the triangle $1 \leq i \leq j \leq n$ onto the linear interval $1 \leq k \leq \binom{n+1}{2}$.

53.5.5. In order to exhibit the performance of the substitution rule in case of nested occurrences of the lambda notation, consider

«**procedure** $aa(x[\mathbf{1, 2}])$; **array** x ;
 begin
  integer i ;
  $\vdots$
  $x[i, j] := \dots$;
  $bb(x[i, \mathbf{1}])$;
 end aa»,

where bb is declared as

«**procedure** $bb(y[\mathbf{1}])$; **array** y ;
 begin
  integer i ;
  $\vdots$
  $y[i] := \dots$
 end bb».

Now the equivalence block for a call «$aa(z[\mathbf{2}, f, \mathbf{1}, -\mathbf{1}])$» is determined in two steps: First, for the call of aa:

«**begin**
 integer i ;
 $\vdots$
 $z[j, f, i, -i] := \dots$;
 $bb(z[\mathbf{1}, f, i, -i])$;
 end aa»,

and then for the call of bb within aa:

«**begin**
 integer i ;
 $\vdots$
 $z[j, f, i, -i] := \dots$;
 begin
  integer $i\alpha$;
  $\vdots$
  $z[i\alpha, f, i, -i] := \dots$;
 end bb
 end aa».

Conversely, we could also use the semistatic rule of 48.2.2 which produces first a new procedure

```
«procedure aa1 (x [1, 2]) ; array x ;
    begin
      integer i ;
        ⋮
      x [i, j] := … ;
      begin
        integer iæ ;
          ⋮
        x [i, iæ] := … ;
      end bb
    end aa1»,
```

whereupon application of the substitution rule given in 53.4 yields the same equivalence block as above.

We conclude with the remark, that there are of course other desirable extensions of the present ALGOL, but this one is important insofar as efficient handling of arrays and subscripted variables is a predominant requirement in computing practice.

Appendix B. The IFIP-Reports on ALGOL

Revised Report on the Algorithmic Language ALGOL 60* [1]

By

J. W. BACKUS, F. L. BAUER, J. GREEN, C. KATZ, J. MCCARTHY,
P. NAUR, A. J. PERLIS, H. RUTISHAUSER, K. SAMELSON, B. VAUQUOIS,
J. H. WEGSTEIN, A. VAN WIJNGAARDEN, M. WOODGER

Edited by

PETER NAUR

Dedicated to the memory of WILLIAM TURANSKI

Summary

The report gives a complete defining description of the international algorithmic language ALGOL 60. This is a language suitable for expressing a large class of numerical processes in a form sufficiently concise for direct automatic translation into the language of programmed automatic computers.

The introduction contains an account of the preparatory work leading up to the final conference, where the language was defined. In addition the notions reference language, publication language, and hardware representations are explained.

In the first chapter a survey of the basic constituents and features of the language is given, and the formal notation, by which the syntactic structure is defined, is explained.

The second chapter lists all the basic symbols, and the syntactic units known as identifiers, numbers, and strings are defined. Further some important notions such as quantity and value are defined.

The third chapter explains the rules for forming expressions and the meaning of these expressions. Three different types of expressions exist: arithmetic, Boolean (logical), and designational.

The fourth chapter describes the operational units of the language, known as statements. The basic statements are: assignment statements (evaluation of a formula), go to statements (explicit break of the sequence

---

* International Federation for Information Processing 1962.
[1] Numerische Mathematik **4**, 420—453 (1963).

of execution of statements), dummy statements, and procedure statements (call for execution of a closed process, defined by a procedure declaration). The formation of more complex structures, having statement character, is explained. These include: conditional statements, for statements, compound statements, and blocks.

In the fifth chapter the units known as declarations, serving for defining permanent properties of the units entering into a process described in the language, are defined.

The report ends with two detailed examples of the use of the language and an alphabetic index of definitions.

Introduction

Background

After the publication[1,2] of a preliminary report on the algorithmic language ALGOL, as prepared at a conference in Zürich in 1958, much interest in the ALGOL language developed.

As a result of an informal meeting held at Mainz in November 1958, about forty interested persons from several European countries held an ALGOL implementation conference in Copenhagen in February 1959. A "hardware group" was formed for working cooperatively right down to the level of the paper tape code. This conference also led to the publication by Regnecentralen, Copenhagen, of an *Algol Bulletin*, edited by PETER NAUR, which served as a forum for further discussion. During the June 1959 ICIP Conference in Paris several meetings, both formal and informal ones, were held. These meetings revealed some misunderstandings as to the intent of the group which was primarily responsible for the formulation of the language, but at the same time made it clear that there exists a wide appreciation of the effort involved. As a result of the discussions it was decided to hold an international meeting in January 1960 for improving the ALGOL language and preparing a final report. At a European ALGOL Conference in Paris in November 1959 which was attended by about fifty people, seven European representatives were selected to attend the January 1960 Conference, and they represent the following organizations: Association Française de Calcul, British Computer Society, Gesellschaft für Angewandte Mathematik und Mechanik, and Nederlands Rekenmachine Genootschap. The seven representatives held a final preparatory meeting at Mainz in December 1959.

---

[1] Preliminary report — International Algebraic Language, Comm. Assoc. Comp. Mach. 1, No. 12 (1958), 8.

[2] Report on the Algorithmic Language ALGOL by the ACM Committee on Programming Languages and the GAMM Committee on Programming, edited by A. J. PERLIS and K. SAMELSON, Numerische Mathematik Bd. 1, S. 41—60 (1959).

Meanwhile, in the United States, anyone who wished to suggest changes or corrections to ALGOL was requested to send his comments to the *Communications of the ACM*, where they were published. These comments then became the basis of consideration for changes in the ALGOL language. Both the SHARE and USE organizations established ALGOL working groups, and both organizations were represented on the ACM Committee on Programming Languages. The ACM Committee met in Washington in November 1959 and considered all comments on ALGOL that had been sent to the ACM *Communications*. Also, seven representatives were selected to attend the January 1960 international conference. These seven representatives held a final preparatory meeting in Boston in December 1959.

January 1960 Conference

The thirteen representatives[1], from Denmark, England, France, Germany, Holland, Switzerland, and the United States, conferred in Paris from January 11 to 16, 1960.

Prior to this meeting a completely new draft report was worked out from the preliminary report and the recommendations of the preparatory meetings by PETER NAUR and the Conference adopted this new form as the basis for its report. The Conference then proceeded to work for agreement on each item of the report. The present report represents the union of the Committee's concepts and the intersection of its agreements.

April 1962 Conference [Edited by M. Woodger]

A meeting of some of the authors of ALGOL 60 was held on 2nd—3rd April 1962 in Rome, Italy, through the facilities and courtesy of the International Computation Centre. The following were present:

Authors	*Advisers*	*Observer*
F. L. BAUER	M. PAUL	W. L. VAN DER POEL
J. GREEN	R. FRANCIOTTI	(Chairman, IFIP TC 2.1
C. KATZ	P. Z. INGERMAN	Working Group ALGOL)
R. KOGON (representing		
J. W. BACKUS)		
P. NAUR		
K. SAMELSON	G. SEEGMÜLLER	
J. H. WEGSTEIN	R. E. UTMAN	
A. VAN WIJNGAARDEN		
M. WOODGER	P. LANDIN	

[1] WILLIAM TURANSKI of the American group was killed by an automobile just prior to the January 1960 Conference.

The purpose of the meeting was to correct known errors in, attempt to eliminate apparent ambiguities in, and otherwise clarify the ALGOL 60 Report. Extensions to the language were not considered at the meeting. Various proposals for correction and clarification that were submitted by interested parties in response to the Questionnaire in *Algol Bulletin* No. 14 were used as a guide.

This report[1] constitutes a supplement to the ALGOL 60 Report which should resolve a number of difficulties therein. Not all of the questions raised concerning the original report could be resolved. Rather than risk hastily drawn conclusions on a number of subtle points, which might create new ambiguities, the committee decided to report only those points which they unanimously felt could be stated in clear and unambiguous fashion.

Questions concerned with the following areas are left for further consideration by Working Group 2.1 of IFIP, in the expectation that current work on advanced programming languages will lead to better resolution:

1. Side effects of functions.
2. The call by name concept.
3. **own**: static or dynamic.
4. For statement: static or dynamic.
5. Conflict between specification and declaration.

The authors of the ALGOL 60 Report present at the Rome Conference, being aware of the formation of a Working Group on ALGOL by IFIP, accepted that any collective responsibility which they might have with respect to the development, specification, and refinement of the ALGOL language will from now on be transferred to that body.

This report has been reviewed by IFIP TC 2 on Programming Languages in August 1962 and has been approved by the Council of the International Federation for Information Processing.

As with the preliminary ALGOL report, three different levels of language are recognized, namely a Reference Language, a Publication Language, and several Hardware Representations.

---

[1] [Editor's note: — The present edition follows the text which was approved by the Council of IFIP. Although it is not clear from the Introduction, the present version is the original report of the January 1960 conference modified according to the agreements reached during the April 1962 conference. Thus the report mentioned here is incorporated in the present version. The modifications touch the original report in the following sections: Changes of text: 1 with footnote; 2.1 footnote; 2.3; 2.7; 3.3.3; 3.3.4.2; 4.1.3; 4.2.3; 4.2.4; 4.3.4; 4.7.3; 4.7.3.1; 4.7.3.3; 4.7.5.1; 4.7.5.4; 4.7.6; 5; 5.3.3; 5.3.5; 5.4.3; 5.4.4; 5.4.5. Changes of syntax: 3.4.1; 4.1.1; 4.2.1; 4.5.1.]

Reference Language

1. It is the working language of the committee.

2. It is the defining language.

3. The characters are determined by ease of mutual understanding and not by any computer limitations, coders notation, or pure mathematical notation.

4. It is the basic reference and guide for compiler builders.

5. It is the guide for all hardware representations.

6. It is the guide for transliterating from publication language to any locally appropriate hardware representations.

7. The main publications of the ALGOL language itself will use the reference representation.

Publication Language

1. The publication language admits variations of the reference language according to usage of printing and handwriting (e.g., subscripts, spaces, exponents, Greek letters).

2. It is used for stating and communicating processes.

3. The characters to be used may be different in different countries, but univocal correspondence with reference representation must be secured.

Hardware Representations

1. Each one of these is a condensation of the reference language enforced by the limited number of characters on standard input equipment.

2. Each one of these uses the character set of a particular computer and is the language accepted by a translator for that computer.

3. Each one of these must be accompanied by a special set of rules for transliterating from publication or reference language.

For transliteration between the reference language and a language suitable for publications, among others, the following rules are recommended.

Reference language	*Publication language*
Subscript brackets []	Lowering of the line between the brackets and removal of the brackets.
Exponentiation ↑	Raising of the exponent.
Parentheses ()	Any form of parentheses, brackets, braces.
Basis of ten $_{10}$	Raising of the ten and of the following integral number, inserting of the intended multiplication sign.

Description of the reference language

> Was sich überhaupt sagen läßt, läßt sich
> klar sagen; und wovon man nicht reden
> kann, darüber muß man schweigen.
>
> LUDWIG WITTGENSTEIN

1. Structure of the language

As stated in the introduction, the algorithmic language has three different kinds of representations — reference, hardware, and publication — and the development described in the sequel is in terms of the reference representation. This means that all objects defined within the language are represented by a given set of symbols — and it is only in the choice of symbols that the other two representations may differ. Structure and content must be the same for all representations.

The purpose of the algorithmic language is to describe computational processes. The basic concept used for the description of calculating rules is the well known arithmetic expression containing as constituents numbers, variables, and functions. From such expressions are compounded, by applying rules of arithmetic composition, self-contained units of the language — explicit formulae — called assignment statements.

To show the flow of computational processes, certain non-arithmetic statements and statement clauses are added which may describe e.g., alternatives, or iterative repetitions of computing statements. Since it is necessary for the function of these statements that one statement refers to another, statements may be provided with labels. A sequence of statements may be enclosed between the statement brackets **begin** and **end** to form a compound statement.

Statements are supported by declarations which are not themselves computing instructions, but inform the translator of the existence and certain properties of objects appearing in statements, such as the class of numbers taken on as values by a variable, the dimension of an array of numbers, or even the set of rules defining a function. A sequence of declarations followed by a sequence of statements and enclosed between **begin** and **end** constitutes a block. Every declaration appears in a block in this way and is valid only for that block.

A program is a block or compound statement which is not contained within another statement and which makes no use of other statements not contained within it.

In the sequel the syntax and semantics of the language will be given[1].

---

[1] Whenever the precision of arithmetic is stated as being in general not specified, or the outcome of a certain process is left undefined or said to be undefined, this is to be interpreted in the sense that a program only fully defines a computational process if the accompanying information specifies the precision assumed, the kind of arithmetic assumed, and the course of action to be taken in all such cases as may occur during the execution of the computation.

1.1. Formalism for syntactic description.

The syntax will be described with the aid of metalinguistic formulae[1]. Their interpretation is best explained by an example:

$$\langle ab \rangle ::= (\mid [\mid \langle ab \rangle (\mid \langle ab \rangle \langle d \rangle$$

Sequences of characters enclosed in the bracket $\langle \rangle$ represent metalinguistic variables whose values are sequences of symbols. The marks ::= and | (the latter with the meaning of **or**) are metalinguistic connectives. Any mark in a formula, which is not a variable or a connective, denotes itself (or the class of marks which are similar to it). Juxta position of marks and/or variables in a formula signifies juxtaposition of the sequences denoted. Thus the formula above gives a recursive rule for the formation of values of the variable $\langle ab \rangle$. It indicates that $\langle ab \rangle$ may have the value (or [or that given some legitimate value of $\langle ab \rangle$, another may be formed by following it with the character (or by following it with some value of the variable $\langle d \rangle$. If the values of $\langle d \rangle$ are the decimal digits, some values of $\langle ab \rangle$ are:

[(((1(37(
(12345(
(((
[86

In order to facilitate the study, the symbols used for distinguishing the metalinguistic variables (i.e. the sequences of characters appearing within the brackets $\langle \rangle$ as ab in the above example) have been chosen to be words describing approximately the nature of the corresponding variable. Where words which have appeared in this manner are used elsewhere in the text they will refer to the corresponding syntactic definition. In addition some formulae have been given in more than one place.

Definition:

$\langle empty \rangle ::=$
(i.e. the null string of symbols).

2. Basic symbols, identifiers, numbers, and strings.

Basic concepts

The reference language is built up from the following basic symbols:

$\langle basic\ symbol \rangle ::= \langle letter \rangle \mid \langle digit \rangle \mid \langle logical\ value \rangle \mid \langle delimiter \rangle$

---

[1] Cf. J. W. BACKUS, The syntax and semantics of the proposed international algebraic language of the Zürich ACM-GAMM conference. ICIP Paris, June 1959.

2.1. Letters

⟨letter⟩ ::= $a|b|c|d|e|f|g|h|i|j|k|l|m|n|o|p|q|r|s|t|u|v|w|x|y|z|$
$A|B|C|D|E|F|G|H|I|J|K|L|M|N|O|P|Q|R|S|T|U|V|W|X|Y|Z$

This alphabet may arbitrarily be restricted, or extended with any other distinctive character (i.e. character not coinciding with any digit, logical value or delimiter).

Letters do not have individual meaning. They are used for forming identifiers and strings[1] (cf. sections 2.4. Identifiers, 2.6. Strings).

2.2.1. Digits

⟨digit⟩ ::= $0|1|2|3|4|5|6|7|8|9$

Digits are used for forming numbers, identifiers, and strings.

2.2.2. Logical values

⟨logical value⟩ ::= **true** | **false**

The logical values have a fixed obvious meaning.

2.3. Delimiters

⟨delimiter⟩ ::= ⟨operator⟩ | ⟨separator⟩ | ⟨bracket⟩ | ⟨declarator⟩ |
    ⟨specificator⟩
⟨operator⟩ ::= ⟨arithmetic operator⟩ | ⟨relational operator⟩ |
    ⟨logical operator⟩ | ⟨sequential operator⟩
⟨arithmetic operator⟩ ::= $+$ | $-$ | $\times$ | $/$ | $\div$ | $\uparrow$
⟨relational operator⟩ ::= $<$ | $\leq$ | $=$ | $\geq$ | $>$ | $\neq$
⟨logical operator⟩ ::= $\equiv$ | $\supset$ | $\vee$ | $\wedge$ | $\neg$
⟨sequential operator⟩ ::= **goto** | **if** | **then** | **else** | **for** | **do** [2]
⟨separator⟩ ::= , | . | $_{10}$ | : | ; | := | ⊔ | **step** | **until** | **while** | **comment**
⟨bracket⟩ ::= (|) |[|]| '|' | **begin** | **end**
⟨declarator⟩ ::= **own** | **Boolean** | **integer** | **real** | **array** | **switch** |
    procedure
⟨specificator⟩ ::= **string** | **label** | **value**

Delimiters have a fixed meaning which for the most part is obvious or else will be given at the appropriate place in the sequel.

---

[1] It should be particularly noted that throughout the reference language underlining [in typewritten copy; boldface type in printed copy — Ed.] is used for defining independent basic symbols (see sections 2.2.2 and 2.3). These are understood to have no relation to the individual letters of which they are composed. Within the present report [not including headings — Ed.] underlining [boldface — Ed.] will be used for no other purposes.

[2] **do** is used in for statements. It has no relation whatsoever to the *do* of the preliminary report, which is not included in ALGOL 60.

Typographical features such as blank space or change to a new line have no significance in the reference language. They may, however, be used freely for facilitating reading.

For the purpose of including text among the symbols of a program the following "comment" conventions hold:

The sequence of basic symbols:	is equivalent to
; **comment** ⟨any sequence not containing ;⟩;	;
begin comment ⟨any sequence not containing ;⟩;	**begin**
end ⟨any sequence not containing **end** or ; or **else**⟩	**end**

By equivalence is here meant that any of the three structures shown in the left hand column may be replaced, in any occurrence outside of strings, by the symbol shown on the same line in the right hand column without any effect on the action of the program. It is further understood that the comment structure encountered first in the text when reading from left to right has precedence in being replaced over later structures contained in the sequence.

2.4. Identifiers

2.4.1. Syntax.

⟨identifier⟩ ::= ⟨letter⟩ | ⟨identifier⟩ ⟨letter⟩ | ⟨identifier⟩ ⟨digit⟩

2.4.2. Examples

$$q$$
$$Soup$$
$$V17a$$
$$a34kTMNs$$
$$MARILYN$$

2.4.3. Semantics. Identifiers have no inherent meaning, but serve for the identification of simple variables, arrays, labels, switches, and procedures. They may be chosen freely (cf. however section 3.2.4. Standard functions).

The same identifier cannot be used to denote two different quantities except when these quantities have disjoint scopes as defined by the declarations of the program (cf. section 2.7. Quantities, kinds and scopes and section 5. Declarations).

2.5. Numbers

2.5.1. Syntax.
⟨unsigned integer⟩ ::= ⟨digit⟩ | ⟨unsigned integer⟩ ⟨digit⟩
⟨integer⟩ ::= ⟨unsigned integer⟩ | + ⟨unsigned integer⟩ |
    − ⟨unsigned integer⟩
⟨decimal fraction⟩ ::= . ⟨unsigned integer⟩
⟨exponent part⟩ ::= $_{10}$ ⟨integer⟩

⟨decimal number⟩ ::= ⟨unsigned integer⟩ | ⟨decimal fraction⟩ |
    ⟨unsigned integer⟩ ⟨decimal fraction⟩
⟨unsigned number⟩ ::= ⟨decimal number⟩ | ⟨exponent part⟩ |
    ⟨decimal number⟩ ⟨exponent part⟩
⟨number⟩ ::= ⟨unsigned number⟩ | + ⟨unsigned number⟩ |
    − ⟨unsigned number⟩

2.5.2. Examples

0	-200.084	$-.083_{10}-02$
177	$+\ 07.43_{10}8$	$-_{10}7$
$.5384$	$9.34_{10}+10$	$_{10}-4$
$+0.7300$	$2_{10}-4$	$+_{10}+5$

2.5.3. Semantics. Decimal numbers have their conventional meaning. The exponent part is a scale factor expressed as an integral power of 10.

2.5.4. Types. Integers are of type **integer**. All other numbers are of type **real** (cf. section 5.1 Type declarations).

2.6. Strings

2.6.1. Syntax.

⟨proper string⟩ ::= ⟨any sequence of basic symbols not containing ' or '⟩|
    ⟨empty⟩
⟨open string⟩ ::= ⟨proper string⟩ | '⟨open string⟩' |
    ⟨open string⟩ ⟨open string⟩
⟨string⟩ ::= '⟨open string⟩'

2.6.2. Examples.   '5k,, − '[[['∧=/:'Tt''
                          '..This ⊔ is ⊔ a ⊔ 'string''

2.6.3. Semantics. In order to enable the language to handle arbitrary sequences of basic symbols the string quotes ' and ' are introduced. The symbol ⊔ denotes a space. It has no significance outside strings.

Strings are used as actual parameters of procedures (cf. sections 3.2. Function designators and 4.7. Procedure statements).

2.7. Quantities, kinds and scopes

The following kinds of quantities are distinguished: simple variables, arrays, labels, switches, and procedures.

The scope of a quantity is the set of statements and expressions in which the declaration of the identifier associated with that quantity is valid. For labels see section 4.1.3.

2.8. Values and types

A value is an ordered set of numbers (special case: a single number), an ordered set of logical values (special case: a single logical value), or a label.

Certain of the syntactic units are said to possess values. These values will in general change during the execution of the program. The values of expressions and their constituents are defined in section 3. The value of an array identifier is the ordered set of values of the corresponding array of subscripted variables (cf. section 3.1.4.1).

The various "types" (**integer, real, Boolean**) basically denote properties of values. The types associated with syntactic units refer to the values of these units.

3. Expressions

In the language the primary constituents of the programs describing algorithmic processes are arithmetic, Boolean, and designational expressions. Constituents of these expressions, except for certain delimiters, are logical values, numbers, variables, function designators, and elementary arithmetic, relational, logical, and sequential operators. Since the syntactic definition of both variables and function designators contains expressions, the definition of expressions, and their constituents, is necessarily recursive.

⟨expression⟩ ::= ⟨arithmetic expression⟩ | ⟨Boolean expression⟩ |
    ⟨designational expression⟩

3.1. Variables

3.1.1. Syntax.
⟨variable identifier⟩ ::= ⟨identifier⟩
⟨simple variable⟩ ::= ⟨variable identifier⟩
⟨subscript expression⟩ ::= ⟨arithmetic expression⟩
⟨subscript list⟩ ::= ⟨subscript expression⟩ | ⟨subscript list⟩,
    ⟨subscript expression⟩
⟨array identifier⟩ ::= ⟨identifier⟩
⟨subscripted variable⟩ ::= ⟨array identifier⟩ [⟨subscript list⟩]
⟨variable⟩ ::= ⟨simple variable⟩ | ⟨subscripted variable⟩

3.1.2. Examples   *epsilon*
                         detA
                         a17
                         Q[7, 2]
                         x[sin(n×pi/2), Q[3, n, 4]]

3.1.3. Semantics. A variable is a designation given to a single value. This value may be used in expressions for forming other values and may be changed at will by means of assignment statements (section 4.2). The type of the value of a particular variable is defined in the declaration for the variable itself (cf. section 5.1. Type declarations) or for the corresponding array identifier (cf. section 5.2. Array declarations).

3.1.4. Subscripts. 3.1.4.1. Subscripted variables designate values which are components of multidimensional arrays (cf. section 5.2. Array declarations). Each arithmetic expression of the subscript list occupies one subscript position of the subscripted variable and is called a subscript. The complete list of subscripts is enclosed in the subscript brackets []. The array component referred to by a subscripted variable is specified by the actual numerical value of its subscripts (cf. section 3.3. Arithmetic expressions).

3.1.4.2. Each subscript position acts like a variable of type **integer** and the evaluation of the subscript is understood to be equivalent to an assignment to this fictitious variable (cf. section 4.2.4). The value of the subscripted variable is defined only if the value of the subscript expression is within the subscript bounds of the array (cf. section 5.2. Array declarations).

3.2. Function designators

3.2.1. Syntax.

⟨procedure identifier⟩ ::= ⟨identifier⟩
⟨actual parameter⟩ ::= ⟨string⟩ | ⟨expression⟩ | ⟨array identifier⟩ |
    ⟨switch identifier⟩ | ⟨procedure identifier⟩
⟨letter string⟩ ::= ⟨letter⟩ | ⟨letter string⟩ ⟨letter⟩
⟨parameter delimiter⟩ ::= , |) ⟨letter string⟩ : (
⟨actual parameter list⟩ ::= ⟨actual parameter⟩ |
    ⟨actual parameter list⟩ ⟨parameter delimiter⟩ ⟨actual parameter⟩
⟨actual parameter part⟩ ::= ⟨empty⟩ | (⟨actual parameter list⟩)
⟨function designator⟩ ::= ⟨procedure identifier⟩
    ⟨actual parameter part⟩

3.2.2. Examples. $sin(a-b)$
                        $J(v+s, n)$
                        R
                        $S(s-5)$ $Temperature: (T)$ $Pressure: (P)$
                        $Compile\ (':=')\ Stack: (Q)$

3.2.3. Semantics. Function designators define single numerical or logical values which result through the application of given sets of rules defined by a procedure declaration (cf. section 5.4. Procedure declarations) to fixed sets of actual parameters. The rules governing specification of actual parameters are given in section 4.7. Procedure statements. Not every procedure declaration defines the value of a function designator.

3.2.4. Standard functions. Certain identifiers should be reserved for the standard functions of analysis, which will be expressed as procedures. It is recommended that this reserved list should contain:

18 Rutishauser, Description of ALGOL 60

abs (E)   for the modulus (absolute value) of the value of the expression E

sign (E)   for the sign of the value of E ($+1$ for $E > 0$, 0 for $E = 0$, -1 for $E < 0$)

sqrt (E)   for the square root of the value of E

sin (E)   for the sine of the value of E

cos (E)   for the cosine of the value of E

arctan (E) for the principal value of the arctangent of the value of E

ln (E)   for the natural logarithm of the value of E

exp (E)   for the exponential function of the value of E (e^E)

These functions are all understood to operate indifferently on arguments both of type **real** and **integer**. They will all yield values of type **real**, except for *sign* (E) which will have values of type **integer**. In a particular representation these functions may be available without explicit declarations (cf. section 5. Declarations).

3.2.5. Transfer functions. It is understood that transfer functions between any pair of quantities and expressions may be defined. Among the standard functions it is recommended that there be one, namely

entier (E),

which "transfers" an expression of real type to one of integer type, and assigns to it the value which is the largest integer not greater than the value of E.

3.3. Arithmetic expressions

3.3.1. Syntax.

⟨adding operator⟩ ::= + | −
⟨multiplying operator⟩ ::= × | / | ÷
⟨primary⟩ ::= ⟨unsigned number⟩ | ⟨variable⟩ | ⟨function designator⟩ |
     (⟨arithmetic expression⟩)
⟨factor⟩ ::= ⟨primary⟩ | ⟨factor⟩ ↑ ⟨primary⟩
⟨term⟩ ::= ⟨factor⟩ | ⟨term⟩ ⟨multiplying operator⟩ ⟨factor⟩
⟨simple arithmetic expression⟩ ::= ⟨term⟩ | ⟨adding operator⟩
     ⟨term⟩ | ⟨simple arithmetic expression⟩ ⟨adding operator⟩ ⟨term⟩
⟨if clause⟩ ::= **if** ⟨Boolean expression⟩ **then**
⟨arithmetic expression⟩ ::= ⟨simple arithmetic expression⟩ |
     ⟨if clause⟩ ⟨simple arithmetic expression⟩ **else**
     ⟨arithmetic expression⟩

3.3.2. Examples.

Primaries:

$7.394_{10} - 8$

sum
$w[i+2, 8]$
$cos(y+z\times 3)$
$(a-3/y+vu\uparrow 8)$

Factors:
omega
$sum \uparrow cos(y+z\times 3)$
$7.394_{10}-8 \uparrow w[i+2, 8] \uparrow (a-3/y+vu \uparrow 8)$

Terms:
U
$omega \times sum \uparrow cos(y+z\times 3)/7.394_{10}-8 \uparrow w[i+2, 8] \uparrow (a-3/y+vu\uparrow 8)$

Simple arithmetic expression:

$U-Yu+omega \times sum \uparrow cos(y+z\times 3)/7.394_{10}-8 \uparrow w[i+2, 8] \uparrow$
$(a-3/y+vu \uparrow 8)$

Arithmetic expressions:

$w\times u-Q(S+Cu)\uparrow 2$
if $q>0$ **then** $S+3\times Q/A$ **else** $2\times S+3\times q$
if $a<0$ **then** $U+V$ **else if** $a\times b>17$ **then** U/V **else if** $k\ne y$ **then**
   V/U **else** 0
$a\times sin(omega\times t)$
$0.57_{10}12\times a[N\times(N-1)/2, 0]$
$(A\times arctan(y)+Z)\uparrow(7+Q)$
if q **then** $n-1$ **else** n
if $a<0$ **then** A/B **else if** $b=0$ **then** B/A **else** z

3.3.3. Semantics. An arithmetic expression is a rule for computing a numerical value. In case of simple arithmetic expressions this value is obtained by executing the indicated arithmetic operations on the actual numerical values of the primaries of the expression, as explained in detail in section 3.3.4 below. The actual numerical value of a primary is obvious in the case of numbers. For variables it is the current value (assigned last in the dynamic sense), and for function designators it is the value arising from the computing rules defining the procedure (cf. section 5.4.4. Values of function designators) when applied to the current values of the procedure parameters given in the expression. Finally, for arithmetic expressions enclosed in parentheses the value must through a recursive analysis be expressed in terms of the values of primaries of the other three kinds.

In the more general arithmetic expressions, which include if clauses, one out of several simple arithmetic expressions is selected on the basis of the actual values of the Boolean expressions (cf. section 3.4. Boolean

expressions). This selection is made as follows: The Boolean expressions
of the if clauses are evaluated one by one in sequence from left to right
until one having the value **true** is found. The value of the arithmetic
expression is then the value of the first arithmetic expression following
this Boolean (the largest arithmetic expression found in this position is
understood). The construction:

> **else** ⟨simple arithmetic expression⟩

is equivalent to the construction:

> **else if true then** ⟨simple arithmetic expression⟩

3.3.4. Operators and types. Apart from the Boolean expressions of if
clauses, the constituents of simple arithmetic expressions must be of
types **real** or **integer** (cf. section 5.1. Type declarations). The meaning
of the basic operators and the types of the expressions to which they lead
are given by the following rules:

3.3.4.1. The operators $+$, $-$, and $\times$ have the conventional meaning
(addition, subtraction, and multiplication). The type of the expression
will be **integer** if both of the operands are of **integer** type, otherwise
real.

3.3.4.2. The operations ⟨term⟩/⟨factor⟩ and ⟨term⟩$\div$⟨factor⟩ both
denote division, to be understood as a multiplication of the term by the
reciprocal of the factor with due regard to the rules of precedence (cf.
section 3.3.5). Thus for example

$$a/b \times 7/(p-q) \times v/s$$

means

$$((((a \times (b^{-1})) \times 7) \times ((p-q)^{-1})) \times v) \times (s^{-1})$$

The operator / is defined for all four combinations of types **real** and
integer and will yield results of **real** type in any case. The operator $\div$
is defined only for two operands both of type **integer** and will yield a
result of type **integer**, mathematically defined as follows:

$$a \div b = sign\,(a/b) \times entier\,(abs\,(a/b))$$

(cf. sections 3.2.4 and 3.2.5).

3.3.4.3. The operation ⟨factor⟩ $\uparrow$ ⟨primary⟩ denotes exponentiation,
where the factor is the base and the primary is the exponent. Thus for
example

$$2 \uparrow n \uparrow k \qquad \text{means} \qquad (2^n)^k$$

while

$$2 \uparrow (n \uparrow m) \qquad \text{means} \qquad 2^{(n^m)}$$

Writing i for a number of **integer** type, r for a number of **real** type, and a for a number of either **integer** or **real** type, the result is given by the following rules:

$a \uparrow i$ If $i > 0$: $a \times a \times \cdots \times a$ (i times), of the same type as a.

         If $i = 0$,   if $a \neq 0$:   *1*, of the same type as a.

                     if $a = 0$:   undefined.

         If $i < 0$,   if $a \neq 0$: $1/(a \times a \times \cdots \times a)$ (the denominator has $-i$ factors), of type **real**.

                     if $a = 0$:   undefined.

$a \uparrow r$ If $a > 0$: $\exp(r \times ln(a))$, of type **real**.

         If $a = 0$,   if $r > 0$:   *0.0*, of type **real**.

                   if $r \leq 0$:   undefined.

         If $a < 0$:   always undefined.

3.3.5. Precedence of operators. The sequence of operations within one expression is generally from left to right, with the following additional rules:

3.3.5.1. According to the syntax given in section 3.3.1 the following rules of precedence hold:

first: $\uparrow$

second: $\times / \div$

third: $+ -$

3.3.5.2. The expression between a left parenthesis and the matching right parenthesis is evaluated by itself and this value is used in subsequent calculations. Consequently the desired order of execution of operations within an expression can always be arranged by appropriate positioning of parentheses.

3.3.6. Arithmetics of **real** quantities. Numbers and variables of type **real** must be interpreted in the sense of numerical analysis, i.e. as entities defined inherently with only a finite accuracy. Similarly, the possibility of the occurrence of a finite deviation from the mathematically defined result in any arithmetic expression is explicitly understood. No exact arithmetic will be specified, however, and it is indeed understood that different hardware representations may evaluate arithmetic expressions differently. The control of the possible consequences of such differences must be carried out by the methods of numerical analysis. This control must be considered a part of the process to be described, and will therefore be expressed in terms of the language itself.

3.4. Boolean expressions

3.4.1. Syntax.

⟨relational operator⟩ ::= < | ≦ | = | ≧ | > | ≠

⟨relation⟩ ::= ⟨simple arithmetic expression⟩ ⟨relational operator⟩
    ⟨simple arithmetic expression⟩

⟨Boolean primary⟩ ::= ⟨logical value⟩ | ⟨variable⟩ |
    ⟨function designator⟩ | ⟨relation⟩ | (⟨Boolean expression⟩)

⟨Boolean secondary⟩ ::= ⟨Boolean primary⟩ | ¬ ⟨Boolean primary⟩

⟨Boolean factor⟩ ::= ⟨Boolean secondary⟩ |
    ⟨Boolean factor⟩ ∧ ⟨Boolean secondary⟩

⟨Boolean term⟩ ::= ⟨Boolean factor⟩ | ⟨Boolean term⟩ ∨
    ⟨Boolean factor⟩

⟨implication⟩ ::= ⟨Boolean term⟩ | ⟨implication⟩ ⊃ ⟨Boolean term⟩

⟨simple Boolean⟩ ::= ⟨implication⟩ | ⟨simple Boolean⟩ ≡ ⟨implication⟩

⟨Boolean expression⟩ ::= ⟨simple Boolean⟩ |
    ⟨if clause⟩ ⟨simple Boolean⟩ **else** ⟨Boolean expression⟩

3.4.2. Examples. $x = -2$

$$Y > V \lor z < q$$
$$a + b > -5 \land z - d > q \uparrow 2$$
$$p \land q \lor x \neq y$$
$$g \equiv \neg a \land b \land \neg c \lor d \lor e \supset \neg f$$
if $k < 1$ **then** $s > w$ **else** $h \leq c$
if if if a **then** b **else** c **then** d **else** f **then** g **else** $h < k$

3.4.3. Semantics. A Boolean expression is a rule for computing a logical value. The principles of evaluation are entirely analogous to those given for arithmetic expressions in section 3.3.3.

3.4.4. Types. Variables and function designators entered as Boolean primaries must be declared **Boolean** (cf. section 5.1. Type declarations and section 5.4.4. Values of function designators).

3.4.5. The operators. Relations take on the value **true** whenever the corresponding relation is satisfied for the expressions involved, otherwise **false**.

The meaning of the logical operators ¬ (not), ∧ (and), ∨ (or), ⊃ (implies), and ≡ (equivalent), is given by the following function table.

$b1$	false	false	true	true
$b2$	false	true	false	true
$\neg b1$	true	true	false	false
$b1 \land b2$	false	false	false	true
$b1 \lor b2$	false	true	true	true
$b1 \supset b2$	true	true	false	true
$b1 \equiv b2$	true	false	false	true

3.4.6. Precedence of operators. The sequence of operations within one expression is generally from left to right, with the following additional rules:

3.4.6.1. According to the syntax given in section 3.4.1 the following rules of precedence hold:

first: arithmetic expressions according to section 3.3.5.
second: $< \leq = \geq > \neq$
third: $\neg$
fourth: $\wedge$
fifth: $\vee$
sixth: $\supset$
seventh: $\equiv$

3.4.6.2. The use of parentheses will be interpreted in the sense given in section 3.3.5.2.

3.5. Designational expressions

3.5.1. Syntax.
⟨label⟩ ::= ⟨identifier⟩ | ⟨unsigned integer⟩
⟨switch identifier⟩ ::= ⟨identifier⟩
⟨switch designator⟩ ::= ⟨switch identifier⟩ [⟨subscript expression⟩]
⟨simple designational expression⟩ ::= ⟨label⟩ | ⟨switch designator⟩ |
    (⟨designational expression⟩)
⟨designational expression⟩ ::= ⟨simple designational expression⟩ |
    ⟨if clause⟩ ⟨simple designational expression⟩ **else**
    ⟨designational expression⟩

3.5.2. Examples. *17*
            p9
            Choose[n—1]
            Town [**if** $y < 0$ **then** *N* **else** $N+1$]
            if $Ab < c$ **then** *17* **else** $q[$**if** $w \leq 0$ **then** *2* **else** $n]$

3.5.3. Semantics. A designational expression is a rule for obtaining a label of a statement (cf. section 4. Statements). Again the principle of the evaluation is entirely analogous to that of arithmetic expressions (section 3.3.3). In the general case the Boolean expressions of the if clauses will select a simple designational expression. If this is a label the desired result is already found. A switch designator refers to the corresponding switch declaration (cf. section 5.3. Switch declarations) and by the actual numerical value of its subscript expression selects one of the designational expressions listed in the switch declaration by counting these from left to right. Since the designational expression thus selected may again be a switch designator this evaluation is obviously a recursive process.

3.5.4. The subscript expression. The evaluation of the subscript expression is analogous to that of subscripted variables (cf. section 3.1.4.2). The value of a switch designator is defined only if the subscript expression assumes one of the positive values $1, 2, 3, \ldots, n$, where n is the number of entries in the switch list.

3.5.5. Unsigned integers as labels. Unsigned integers used as labels have the property that leading zeroes do not affect their meaning, e.g. *00217* denotes the same label as *217*.

4. Statements

The units of operation within the language are called statements. They will normally be executed consecutively as written. However, this sequence of operations may be broken by go to statements, which define their successor explicitly, and shortened by conditional statements, which may cause certain statements to be skipped.

In order to make it possible to define a specific dynamic succession, statements may be provided with labels.

Since sequences of statements may be grouped together into compound statements and blocks the definition of statement must necessarily be recursive. Also since declarations, described in section 5, enter fundamentally into the syntactic structure, the syntactic definition of statements must suppose declarations to be already defined.

4.1. Compound statements and blocks

4.1.1. Syntax.
⟨unlabelled basic statement⟩ ::= ⟨assignment statement⟩ |
    ⟨go to statement⟩ | ⟨dummy statement⟩ | ⟨procedure statement⟩
⟨basic statement⟩ ::= ⟨unlabelled basic statement⟩ |
    ⟨label⟩: ⟨basic statement⟩
⟨unconditional statement⟩ ::= ⟨basic statement⟩ |
    ⟨compound statement⟩ | ⟨block⟩
⟨statement⟩ ::= ⟨unconditional statement⟩ | ⟨conditional statement⟩ |
    ⟨for statement⟩
⟨compound tail⟩ ::= ⟨statement⟩ **end** | ⟨statement⟩; ⟨compound tail⟩
⟨block head⟩ ::= **begin** ⟨declaration⟩ | ⟨block head⟩; ⟨declaration⟩
⟨unlabelled compound⟩ ::= **begin** ⟨compound tail⟩
⟨unlabelled block⟩ ::= ⟨block head⟩; ⟨compound tail⟩
⟨compound statement⟩ ::= ⟨unlabelled compound⟩ |
    ⟨label⟩: ⟨compound statement⟩
⟨block⟩ ::= ⟨unlabelled block⟩ | ⟨label⟩: ⟨block⟩
⟨program⟩ ::= ⟨block⟩ | ⟨compound statement⟩

This syntax may be illustrated as follows: Denoting arbitrary statements, declarations, and labels, by the letters S, D, and L, respectively, the basic syntactic units take the forms:

Compound statement:

  L: L: ... **begin** S; S; ... S; S **end**

Block:

  L: L: ... **begin** D; D; .. D; S; S; ... S; S **end**

It should be kept in mind that each of the statements S may again be a complete compound statement or block.

4.1.2. Examples.

  Basic statements:

  $a := p+q$
  goto *Naples*
  Start: *Continue*: $W := 7.993$

Compound statement:

  begin $x := 0$; **for** $y := 1$ **step** 1 **until** n **do** $x := x+A[y]$;
      if $x>q$ **then goto** $STOP$ **else if** $x>w—2$ **then goto** S;
      Aw: St: $W := x+bob$ **end**

Block:

  Q: **begin integer** i, k; **real** w;
      for $i := 1$ **step** 1 **until** m **do**
        for $k := i+1$ **step** 1 **until** m **do**
        begin $w := A[i, k]$;
            $A[i, k] := A[k, i]$;
            $A[k, i] := w$ **end** *for i and k*
  end *block Q*

4.1.3. Semantics. Every block automatically introduces a new level of nomenclature. This is realized as follows: Any identifier occurring within the block may through a suitable declaration (cf. section 5. Declarations) be specified to be local to the block in question. This means (a) that the entity represented by this identifier inside the block has no existence outside it and (b) that any entity represented by this identifier outside the block is completely inaccessible inside the block.

Identifiers (except those representing labels) occurring within a block and not being declared to this block will be non-local to it, i.e. will represent the same entity inside the block and in the level immediately outside it. A label separated by a colon from a statement, i.e. labelling that statement, behaves as though declared in the head of the smallest embracing block, i.e. the smallest block whose brackets **begin** and **end** enclose that statement. In this context a procedure body must be considered as if it were enclosed by **begin** and **end** and treated as a block.

Since a statement of a block may again itself be a block the concepts local and non-local to a block must be understood recursively. Thus an identifier, which is non-local to a block A, may or may not be non-local to the block B in which A is one statement.

4.2. Assignment statements

4.2.1. Syntax.

⟨left part⟩ ::= ⟨variable⟩ := | ⟨procedure identifier⟩ :=
⟨left part list⟩ ::= ⟨left part⟩ | ⟨left part list⟩ ⟨left part⟩
⟨assignment statement⟩ ::= ⟨left part list⟩ ⟨arithmetic expression⟩ |
   ⟨left part list⟩ ⟨Boolean expression⟩

4.2.2. Examples. $s := p[0] := n := n+1+s$
$$n := n+1$$
$$A := B/C-v-q\times S$$
$$S[v, k+2] := 3-arctan(s\times zeta)$$
$$V := Q > Y \wedge Z$$

4.2.3. Semantics. Assignment statements serve for assigning the value of an expression to one or several variables or procedure identifiers. Assignment to a procedure identifier may only occur within the body of a procedure defining the value of a function designator (cf. section 5.4.4). The process will in the general case be understood to take place in three steps as follows:

4.2.3.1. Any subscript expressions occurring in the left part variables are evaluated in sequence from left to right.

4.2.3.2. The expression of the statement is evaluated.

4.2.3.3. The value of the expression is assigned to all the left part variables, with any subscript expressions having values as evaluated in step 4.2.3.1.

4.2.4. Types. The type associated with all variables and procedure identifiers of a left part list must be the same. If this type is **Boolean**, the expression must likewise be **Boolean**. If the type is **real** or **integer**, the expression must be arithmetic. If the type of the arithmetic expression differs from that associated with the variables and procedure identifiers, appropriate transfer functions are understood to be automatically invoked. For transfer from **real** to **integer** type the transfer function is understood to yield a result equivalent to

   $entier(E+0.5)$

where E is the value of the expression. The type associated with a procedure identifier is given by the declarator which appears as the first symbol of the corresponding procedure declaration (cf. section 5.4.4).

4.3. Go to statements

4.3.1. Syntax.

⟨go to statement⟩ ::= **goto** ⟨designational expression⟩

4.3.2. Examples. **goto** *8*

       goto *exit*[$n+1$]

       goto *Town* [**if** $y<0$ **then** N **else** $N+1$]

       goto if $Ab<c$ **then** *17* **else** q [**if** $w<0$ **then** *2* **else** n]

4.3.3. Semantics. A go to statement interrupts the normal sequence of operations, defined by the write-up of statements, by defining its successor explicitly by the value of a designational expression. Thus the next statement to be executed will be the one having this value as its label.

4.3.4. Restriction. Since labels are inherently local, no go to statement can lead from outside into a block. A go to statement may, however, lead from outside into a compound statement.

4.3.5. Go to an undefined switch designator. A go to statement is equivalent to a dummy statement if the designational expression is a switch designator whose value is undefined.

4.4. Dummy statements

4.4.1. Syntax.

⟨dummy statement⟩ ::= ⟨empty⟩

4.4.2. Examples.

  L:

  begin ….; *John*: **end**

4.4.3. Semantics. A dummy statement executes no operation. It may serve to place a label.

4.5. Conditional statements

4.5.1. Syntax.

⟨if clause⟩ ::= **if** ⟨Boolean expression⟩ **then**

⟨unconditional statement⟩ ::= ⟨basic statement⟩ |

  ⟨compound statement⟩ | ⟨block⟩

⟨if statement⟩ ::= ⟨if clause⟩ ⟨unconditional statement⟩

⟨conditional statement⟩ ::= ⟨if statement⟩ |

  ⟨if statement⟩ **else** ⟨statement⟩ | ⟨if clause⟩ ⟨for statement⟩ |

  ⟨label⟩: ⟨conditional statement⟩

4.5.2. Examples.

  if $x>0$ **then** $n:=n+1$

  if $v>u$ **then** $V:q:=n+m$ **else goto** R

  if $s<0$ ∨ $P\leqq Q$ **then** AA: **begin if** $q<v$ **then** $a:=v/s$

          else $y:=2\times a$ **end else if** $v>s$ **then** $a:=v-q$

          else if $v>s-1$ **then goto** S

4.5.3. Semantics. Conditional statements cause certain statements to be executed or skipped depending on the running values of specified Boolean expressions.

4.5.3.1. If statement. The unconditional statement of an if statement will be executed if the Boolean expression of the if clause is true. Otherwise it will be skipped and the operation will be continued with the next statement.

4.5.3.2. Conditional statement. According to the syntax two different forms of conditional statements are possible. These may be illustrated as follows:

    if B1 **then** S1 **else if** B2 **then** S2 **else** S3; S4

and

    if B1 **then** S1 **else if** B2 **then** S2 **else if** B3 **then** S3; S4

Here B1 to B3 are Boolean expressions, while S1 to S3 are unconditional statements. S4 is the statement following the complete conditional statement.

The execution of a conditional statement may be described as follows: The Boolean expressions of the if clauses are evaluated one after the other in sequence from left to right until one yielding the value **true** is found. Then the unconditional statement following this Boolean is executed. Unless this statement defines its successor explicitly the next statement to be executed will be S4, i.e. the statement following the complete conditional statement. Thus the effect of the delimiter **else** may be described by saying that it defines the successor of the statement it follows to be the statement following the complete conditional statement.

The construction

    else ⟨unconditional statement⟩

is equivalent to

    else if true then ⟨unconditional statement⟩

If none of the Boolean expressions of the if clauses is true, the effect of the whole conditional statement will be equivalent to that of a dummy statement.

For further explanation the following picture may be useful:

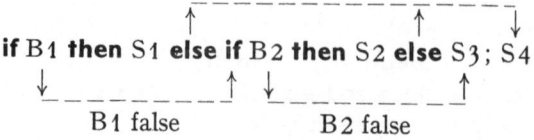

       B1 false   B2 false

4.5.4. Go to into a conditional statement. The effect of a go to statement leading into a conditional statement follows directly from the above explanation of the effect of **else**.

4.6. For statements

4.6.1. Syntax.

⟨for list element⟩ ::= ⟨arithmetic expression⟩ |
 ⟨arithmetic expression⟩ **step** ⟨arithmetic expression⟩ **until**
 ⟨arithmetic expression⟩ |
 ⟨arithmetic expression⟩ **while** ⟨Boolean expression⟩
⟨for list⟩ ::= ⟨for list element⟩ | ⟨for list⟩, ⟨for list element⟩
⟨for clause⟩ ::= **for** ⟨variable⟩ := ⟨for list⟩ **do**
⟨for statement⟩ ::= ⟨for clause⟩ ⟨statement⟩ |
 ⟨label⟩: ⟨for statement⟩

4.6.2. Examples.

> **for** $q := 1$ **step** s **until** n **do** $A[q] := B[q]$
> **for** $k := 1, V1 \times 2$ **while** $V1 < N$ **do**
> **for** $j := I + G, L, 1$ **step** 1 **until** $N, C + D$ **do** $A[k, j] := B[k, j]$

4.6.3. Semantics. A for clause causes the statement S which it precedes to be repeatedly executed zero or more times. In addition it performs a sequence of assignments to its controlled variable. The process may be visualized by means of the following picture:

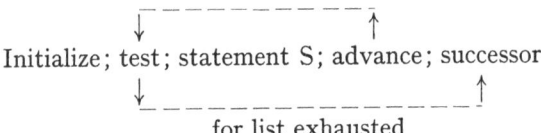

Initialize; test; statement S; advance; successor

for list exhausted

In this picture the word initialize means: perform the first assignment of the for clause. Advance means: perform the next assignment of the for clause. Test determines if the last assignment has been done. If so, the execution continues with the successor of the for statement. If not, the statement following the for clause is executed.

4.6.4. The for list elements. The for list gives a rule for obtaining the values which are consecutively assigned to the controlled variable. This sequence of values is obtained from the for list elements by taking these one by one in the order in which they are written. The sequence of values generated by each of the three species of for list elements and the corresponding execution of the statement S are given by the following rules:

4.6.4.1. Arithmetic expression. This element gives rise to one value, namely the value of the given arithmetic expression as calculated immediately before the corresponding execution of the statement S.

4.6.4.2. Step-until-element. An element of the form A **step** B **until** C, where A, B, and C are arithmetic expressions, gives rise to an execution which may be described most concisely in terms of additional ALGOL statements as follows:

$$V := A;$$
$L1:$ **if** $(V-C) \times \text{sign}(B) > 0$ **then goto** *Element exhausted*;
Statement S;
$$V := V+B;$$
goto $L1$;

where V is the controlled variable of the for clause and *Element exhausted* points to the evaluation according to the next element in the for list, or if the step-until-element is the last of the list, to the next statement in the program.

4.6.4.3. While-element. The execution governed by a for list element of the form E **while** F, where E is an arithmetic and F a Boolean expresion, is most concisely described in terms of additional ALGOL statements as follows:

$L3: V := E;$
if $\neg$ F **then goto** *Element exhausted*;
Statement S;
goto $L3$;

where the notation is the same as in 4.6.4.2 above.

4.6.5. The value of the controlled variable upon exit. Upon exit out of the statement S (supposed to be compound) through a go to statement the value of the controlled variable will be the same as it was immediately preceding the execution of the go to statement.

If the exit is due to exhaustion of the for list, on the other hand, the value of the controlled variable is undefined after the exit.

4.6.6. Go to leading into a for statement

The effect of a go to statement, outside a for statement, which refers to a label within the for statement, is undefined.

4.7. Procedure statements

4.7.1. Syntax.
⟨actual parameter⟩ ::= ⟨string⟩ | ⟨expression⟩ | ⟨array identifier⟩ |
    ⟨switch identifier⟩ | ⟨procedure identifier⟩
⟨letter string⟩ ::= ⟨letter⟩ | ⟨letter string⟩ ⟨letter⟩
⟨parameter delimiter⟩ ::= , |) ⟨letter string⟩: (
⟨actual parameter list⟩ ::= ⟨actual parameter⟩ |
    ⟨actual parameter list⟩ ⟨parameter delimiter⟩ ⟨actual parameter⟩

⟨actual parameter part⟩ ::= ⟨empty⟩ | (⟨actual parameter list⟩)
⟨procedure statement⟩ ::= ⟨procedure identifier⟩
    ⟨actual parameter part⟩

4.7.2. Examples. *Spur (A) Order: (7) Result to: (V)*
                  Transpose (W, v+1)
                  Absmax (A, N, M, Yy, I, K)
                  Innerproduct (A[t, P, u], B[P], 10, P, Y)

These examples correspond to examples given in section 5.4.2.

4.7.3. Semantics. A procedure statement serves to invoke (call for) the execution of a procedure body (cf. section 5.4. procedure declarations). Where the procedure body is a statement written in ALGOL the effect of this execution will be equivalent to the effect of performing the following operations on the program at the time of execution of the procedure statement:

4.7.3.1. Value assignment (call by value). All formal parameters quoted in the value part of the procedure declaration heading are assigned the values (cf. section 2.8. Values and types) of the corresponding actual parameters, these assignments being considered as being performed explicitly before entering the procedure body. The effect is as though an additional block embracing the procedure body were created in which these assignments were made to variables local to this fictitious block with types as given in the corresponding specifications (cf. section 5.4.5). As a consequence, variables called by value are to be considered as non-local to the body of the procedure, but local to the fictitious block (cf. section 5.4.3).

4.7.3.2. Name replacement (call by name). Any formal parameter not quoted in the value list is replaced, throughout the procedure body, by the corresponding actual parameter, after enclosing this latter in parentheses wherever syntactically possible. Possible conflicts between identifiers inserted through this process and other identifiers already present within the procedure body will be avoided by suitable systematic changes of the formal or local identifiers involved.

4.7.3.3. Body replacement and execution. Finally the procedure body, modified as above, is inserted in place of the procedure statement and executed. If the procedure is called from a place outside the scope of any non-local quantity of the procedure body the conflicts between the identifiers inserted through this process of body replacement and the identifiers whose declarations are valid at the place of the procedure statement or function designator will be avoided through suitable systematic changes of the latter identifiers.

4.7.4. Actual-formal correspondence. The correspondence between the actual parameters of the procedure statement and the formal parameters of the procedure heading is established as follows: The actual parameter list of the procedure statement must have the same number of entries as the formal parameter list of the procedure declaration heading. The correspondence is obtained by taking the entries of these two lists in the same order.

4.7.5. Restrictions. For a procedure statement to be defined it is evidently necessary that the operations on the procedure body defined in sections 4.7.3.1 and 4.7.3.2 lead to a correct ALGOL statement.

This imposes the restriction on any procedure statement that the kind and type of each actual parameter be compatible with the kind and type of the corresponding formal parameter. Some important particular cases of this general rule are the following:

4.7.5.1. If a string is supplied as an actual parameter in a procedure statement or function designator, whose defining procedure body is an ALGOL 60 statement (as opposed to non-ALGOL code, cf. section 4.7.8), then this string can only be used within the procedure body as an actual parameter in further procedure calls. Ultimately it can only be used by a procedure body expressed in non-ALGOL code.

4.7.5.2. A formal parameter which occurs as a left part variable in an assignment statement within the procedure body and which is not called by value can only correspond to an actual parameter which is a variable (special case of expression).

4.7.5.3. A formal parameter which is used within the procedure body as an array identifier can only correspond to an actual parameter which is an array identifier of an array of the same dimensions. In addition if the formal parameter is called by value the local array created during the call will have the same subscript bounds as the actual array.

4.7.5.4. A formal parameter which is called by value cannot in general correspond to a switch identifier or a procedure identifier or a string, because these latter do not possess values (the exception is the procedure identifier of a procedure declaration which has an empty formal parameter part (cf. section 5.4.1) and which defines the value of a function designator (cf. section 5.4.4). This procedure identifier is in itself a complete expression).

4.7.5.5. Any formal parameter may have restrictions on the type of the corresponding actual parameter associated with it (these restrictions may, or may not, be given through specifications in the procedure heading). In the procedure statement such restrictions must evidently be observed.

4.7.6. Deleted.

4.7.7. Parameter delimiters. All parameter delimiters are understood to be equivalent. No correspondence between the parameter delimiters used in a procedure statement and those used in the procedure heading is expected beyond their number being the same. Thus the information conveyed by using the elaborate ones is entirely optional.

4.7.8. Procedure body expressed in code. The restrictions imposed on a procedure statement calling a procedure having its body expressed in non-ALGOL code evidently can only be derived from the characteristics of the code used and the intent of the user and thus fall outside the scope of the reference language.

5. Declarations

Declarations serve to define certain properties of the quantities used in the program, and to associate them with identifiers. A declaration of an identifier is valid for one block. Outside this block the particular identifier may be used for other purposes (cf. section 4.1.3).

Dynamically this implies the following: at the time of an entry into a block (through the **begin** since the labels inside are local and therefore inaccessible from outside) all identifiers declared for the block assume the significance implied by the nature of the declarations given. If these identifiers had already been defined by other declarations outside they are for the time being given a new significance. Identifiers which are not declared for the block, on the other hand, retain their old meaning.

At the time of an exit from a block (through **end**, or by a go to statement) all identifiers which are declared for the block lose their local significance.

A declaration may be marked with the additional declarator **own**. This has the following effect: upon a reentry into the block, the values of own quantities will be unchanged from their values at the last exit, while the values of declared variables which are not marked as own are undefined. Apart from labels and formal parameters of procedure declarations and with the possible exception of those for standard functions (cf. sections 3.2.4 and 3.2.5) all identifiers of a program must be declared. No identifier may be declared more than once in any one block head.

Syntax.

⟨declaration⟩ ::= ⟨type declaration⟩ | ⟨array declaration⟩ |
              ⟨switch declaration⟩ | ⟨procedure declaration⟩

5.1. Type declarations

5.1.1. Syntax.
⟨type list⟩ ::= ⟨simple variable⟩ | ⟨simple variable⟩, ⟨type list⟩
⟨type⟩ ::= **real** | **integer** | **Boolean**
⟨local or own type⟩ ::= ⟨type⟩ | **own** ⟨type⟩
⟨type declaration⟩ ::= ⟨local or own type⟩ ⟨type list⟩

5.1.2. Examples. **integer** p, q, s
       own Boolean $Acryl$, n

5.1.3. Semantics. Type declarations serve to declare certain identifiers to
represent simple variables of a given type. Real declared variables may
only assume positive or negative values including zero. Integer declared
variables may only assume positive and negative integral values including
zero. Boolean declared variables may only assume the values **true** and
false.

In arithmetic expressions any position which can be occupied by a
real declared variable may be occupied by an integer declared variable.

For the semantics of **own**, see the fourth paragraph of section 5
above.

5.2. Array declarations

5.2.1. Syntax.
⟨lower bound⟩ ::= ⟨arithmetic expression⟩
⟨upper bound⟩ ::= ⟨arithmetic expression⟩
⟨bound pair⟩ ::= ⟨lower bound⟩ : ⟨upper bound⟩
⟨bound pair list⟩ ::= ⟨bound pair⟩ | ⟨bound pair list⟩, ⟨bound pair⟩
⟨array segment⟩ ::= ⟨array identifier⟩ [⟨bound pair list⟩] |
  ⟨array identifier⟩, ⟨array segment⟩
⟨array list⟩ ::= ⟨array segment⟩ | ⟨array list⟩, ⟨array segment⟩
⟨array declaration⟩ ::= **array** ⟨array list⟩ |
  ⟨local or own type⟩ **array** ⟨array list⟩

5.2.2. Examples. **array** a, b, c $[7: n,\ 2: m]$, s $[-2: 10]$
       own integer array A [**if** $c < 0$ **then** 2 **else** $1: 20$]
       real array q $[-7: -1]$

5.2.3. Semantics. An array declaration declares one or several identifiers
to represent multidimensional arrays of subscripted variables and gives
the dimensions of the arrays, the bounds of the subscripts, and the types
of the variables.

5.2.3.1. Subscript bounds. The subscript bounds for any array are given
in the first subscript bracket following the identifier of this array in the
form of a bound pair list. Each item of this list gives the lower and upper
bound of a subscript in the form of two arithmetic expressions separated

by the delimiter :. The bound pair list gives the bounds of all subscripts taken in order from left to right.

5.2.3.2. Dimensions. The dimensions are given as the number of entries in the bound pair lists.

5.2.3.3. Types. All arrays declared in one declaration are of the same quoted type. If no type declarator is given the type **real** is understood.

5.2.4. Lower upper bound expressions.

5.2.4.1. The expressions will be evaluated in the same way as subscript expressions (cf. section 3.1.4.2).

5.2.4.2. The expressions can only depend on variables and procedures which are non-local to the block for which the array declaration is valid. Consequently in the outermost block of a program only array declarations with constant bounds may be declared.

5.2.4.3. An array is defined only when the values of all upper subscript bounds are not smaller than those of the corresponding lower bounds.

5.2.4.4. The expressions will be evaluated once at each entrance into the block.

5.2.5. The identity of subscripted variables. The identity of a subscripted variable is not related to the subscript bounds given in the array declaration. However, even if an array is declared **own** the values of the corresponding subscripted variables will, at any time, be defined only for those of these variables which have subscripts within the most recently calculated subscript bounds.

5.3. Switch declarations

5.3.1. Syntax.
⟨switch list⟩ ::= ⟨designational expression⟩ |
          ⟨switch list⟩, ⟨designational expression⟩
⟨switch declaration⟩ ::= **switch** ⟨switch identifier⟩ := ⟨switch list⟩

5.3.2. Examples. **switch** S := S1, S2, Q[m], **if** v > −5 **then** S3 **else** S4
          switch Q := p1, w

5.3.3. Semantics. A switch declaration defines the set of values of the corresponding switch designators. These values are given one by one as the values of the designational expressions entered in the switch list. With each of these designational expressions there is associated a positive integer, 1, 2, ..., obtained by counting the items in the list from left to right. The value of the switch designator corresponding to a given value of the subscript expression (cf. section 3.5. Designational expressions) is the value of the designational expression in the switch list having this given value as its associated integer.

5.3.4. Evaluation of expressions in the switch list. An expression in the switch list will be evaluated every time the item of the list in which the expression occurs is referred to, using the current values of all variables involved.

5.3.5. Influence of scopes. If a switch designator occurs outside the scope of a quantity entering into a designational expression in the switch list, and an evaluation of this switch designator selects this designational expression, then the conflicts between the identifiers for the quantities in this expression and the identifiers whose declarations are valid at the place of the switch designator will be avoided through suitable systematic changes of the latter identifiers.

5.4. Procedure declarations

5.4.1. Syntax.

⟨formal parameter⟩ ::= ⟨identifier⟩
⟨formal parameter list⟩ ::= ⟨formal parameter⟩ |
    ⟨formal parameter list⟩ ⟨parameter delimiter⟩ ⟨formal parameter⟩
⟨formal parameter part⟩ ::= ⟨empty⟩ | (⟨formal parameter list⟩)
⟨identifier list⟩ ::= ⟨identifier⟩ | ⟨identifier list⟩, ⟨identifier⟩
⟨value part⟩ ::= **value** ⟨identifier list⟩; | ⟨empty⟩
⟨specifier⟩ ::= **string** | ⟨type⟩ | **array** | ⟨type⟩ **array** | **label** | **switch** |
    procedure | ⟨type⟩ **procedure**
⟨specification part⟩ ::= ⟨empty⟩ | ⟨specifier⟩ ⟨identifier list⟩; |
    ⟨specification part⟩ ⟨specifier⟩ ⟨identifier list⟩;
⟨procedure heading⟩ ::= ⟨procedure identifier⟩ ⟨formal parameter part⟩;
    ⟨value part⟩ ⟨specification part⟩
⟨procedure body⟩ ::= ⟨statement⟩ | ⟨code⟩
⟨procedure declaration⟩ ::=
    procedure ⟨procedure heading⟩ ⟨procedure body⟩ |
    ⟨type⟩ **procedure** ⟨procedure heading⟩ ⟨procedure body⟩

5.4.2. Examples (see also the examples at the end of the report).

```
procedure Spur (a) Order: (n) Result: (s); value n;
array a; integer n; real s;
begin integer k;
s := 0;
for k := 1 step 1 until n do s := s+a[k, k]
end
```

```
procedure Transpose (a) Order: (n); value n;
array a; integer n;
begin real w; integer i, k;
for i := 1 step 1 until n do
    for k := 1+i step 1 until n do
```

```
begin w := a[i, k];
    a[i, k] := a[k, i];
    a[k, i] := w
end
end Transpose
```

```
integer procedure Step (u); real u;
Step := if 0≦u ∧ u≦1 then 1 else 0
```

```
procedure Absmax (a) Size: (n, m) Result: (y) Subscripts: (i, k);
comment The absolute greatest element of the matrix a, of size n by m
is transferred to y, and the subscripts of this element to i and k;
array a; integer n, m, i, k; real y;
begin integer p, q;
y := 0;
for p := 1 step 1 until n do for q := 1 step 1 until m do
if abs (a[p, q]) > y then begin y := abs (a[p, q]);
    i := p; k := q end end Absmax
```

```
procedure    Innerproduct (a, b) Order: (k, p) Result: (y); value k;
integer k, p; real y, a, b;
begin real s;
s := 0;
for p := 1 step 1 until k do s := s+a×b;
y := s
end Innerproduct
```

5.4.3. Semantics. A procedure declaration serves to define the procedure associated with a procedure identifier. The principal constituent of a procedure declaration is a statement or a piece of code, the procedure body, which through the use of procedure statements and/or function designators may be activated from other parts of the block in the head of which the procedure declaration appears. Associated with the body is a heading, which specifies certain identifiers occurring within the body to represent formal parameters. Formal parameters in the procedure body will, whenever the procedure is activated (cf. section 3.2. Function designators and section 4.7. Procedure statements) be assigned the values of or replaced by actual parameters. Identifiers in the procedure body which are not formal will be either local or non-local to the body depending on whether they are declared within the body or not. Those of them which are non-local to the body may well be local to the block in the head of which the procedure declaration appears. The procedure body always acts like a block, whether it has the form of one or not. Consequently the scope of any label labelling a statement within the body or the body itself can never extend beyond the procedure body. In addition, if the identifier of a formal parameter is declared anew within the procedure body (including the case of its use as a label as in section 4.1.3), it is thereby given a local significance and actual parameters which correspond to it are inaccessible throughout the scope of this inner local quantity.

5.4.4. Values of function designators. For a procedure declaration to define the value of a function designator there must, within the procedure body, occur one or more explicit assignment statements with the procedure identifier in a left part; at least one of these must be executed, and the type associated with the procedure identifier must be declared through the appearance of a type declarator as the very first symbol of the procedure declaration. The last value so assigned is used to continue the evaluation of the expression in which the function designator occurs. Any occurrence of the procedure identifier within the body of the procedure other than in a left part in an assignment statement denotes activation of the procedure.

5.4.5. Specifications. In the heading a specification part, giving information about the kinds and types of the formal parameters by means of an obvious notation, may be included. In this part no formal parameter may occur more than once. Specifications of formal parameters called by value (cf. section 4.7.3.1) must be supplied and specifications of formal parameters called by name (cf. section 4.7.3.2) may be omitted.

5.4.6. Code as procedure body. It is understood that the procedure body may be expressed in non-ALGOL language. Since it is intended that the use of this feature should be entirely a question of hardware representation, no further rules concerning this code language can be given within the reference language.

Examples of procedure declarations

Example 1

procedure *euler* (*fct, sum, eps, tim*); **value** *eps, tim*; **integer** *tim*;
real procedure *fct*; **real** *sum, eps*;
comment *euler computes the sum of fct (i) for i from zero up to infinity by means of a suitably refined euler transformation. The summation is stopped as soon as tim times in succession the absolute value of the terms of the transformed series are found to be less than eps. Hence, one should provide a function fct with one integer argument, an upper bound eps, and an integer tim. The output is the sum sum. euler is particularly efficient in the case of a slowly convergent or divergent alternating series;*

```
begin integer i, k, n, t; array m[0: 15]; real mn, mp, ds;
i := n := t := 0; m[0] := fct(0); sum := m[0]/2;
nextterm: i := i+1; mn := fct(i);
        for k := 0 step 1 until n do
            begin mp := (mn+m[k])/2; m[k] := mn;
                mn := mp end means;
        if (abs(mn) < abs(m[n])) ∧ (n < 15) then
            begin ds := mn/2; n := n+1; m[n] := mn end accept
        else ds := mn;
        sum := sum+ds;
        if abs(ds) < eps then t := t+1 else t := 0;
        if t < tim then goto nextterm
end euler
```

Example 2[1]

```
procedure RK(x, y, n, FKT, eps, eta, xE, yE, fi); value x, y; integer n;
Boolean fi; real x, eps, eta, xE; array y, yE; procedure FKT;
comment RK integrates the system y'ₖ = fₖ(x, y₁, y₂, ..., yₙ) (k = 1, 2, ... n)
```

$$y'_k = f_k(x, y_1, y_2, \ldots, y_n) \quad (k = 1, 2, \ldots n)$$

of differential equations with the method of Runge-Kutta with automatic search for appropriate length of integration step. Parameters are: The initial values x and y[k] for x and the unknown functions $y_k(x)$. The order n of the system. The procedure FKT(x, y, n, z) which represents the system to be integrated, i.e. the set of functions f_k. The tolerance values eps and eta which govern the accuracy of the numerical integration. The end of the integration interval xE. The output parameter yE which represents the solution at $x = xE$. The Boolean variable fi, which must always be given the value **true** *for an isolated or first entry into RK. If however the functions y must be available at several meshpoints $x_0, x_1, \ldots, x_n$, then the procedure must be called repeatedly (with $x = x_k$, $xE = x_{k+1}$, for $k = 0, 1, \ldots, n-1$) and then the later calls may occur with fi =* **false** *which saves computing time. The input parameters of FKT must be x, y, n, the output parameter z represents the set of derivatives $z[k] = f_k(x, y[1], y[2], \ldots, y[n])$ for x and the actual y's. A procedure comp enters as a non-local identifier;*

```
begin
    array z, y1, y2, y3[1: n]; real x1, x2, x3, H; Boolean out;
    integer k, j; own real s, Hs;
```

---

[1] This RK-program contains some new ideas which are related to ideas of S. GILL, A process for the step by step integration of differential equations in an automatic computing machine. Proc. Camb. Phil. Soc. **47** (1951) p. 96, and E. FRÖBERG, On the solution of ordinary differential equations with digital computing machines, Fysiograf. Sällsk. Lund, Förhd. **20** Nr. 11 (1950) p. 136—152. It must be clear however that with respect to computing time and round-off errors it may not be optimal, nor has it actually been tested on a computer.

```
procedure RK1ST (x, y, h, xe, ye); real x, h, xe; array y, ye;
    comment RK1ST integrates one single Runge-Kutta step
    with initial values x, y[k] which yields the output parameters
    xe= x+h and ye[k], the latter being the solution at xe.
    Important: the parameters n, FKT, z enter RK1ST as non-
    local entities;
    begin
        array w[1: n], a[1: 5]; integer k, j;
        a[1] := a[2] := a[5] := h/2; a[3] := a[4] := h;
        xe := x;
        for k := 1 step 1 until n do ye[k] := w[k] := y[k];
        for j := 1 step 1 until 4 do
        begin
            FKT(xe, w, n, z);
            xe := x+a[j];
            for k := 1 step 1 until n do
            begin
                w[k] := y[k]+a[j]×z[k];
                ye[k] := ye[k]+a[j+1]×z[k]/3
            end k
        end j
    end RK1ST;
```

Begin of program:

```
    if fi then begin H := xE—x; s := 0 end else H := Hs;
    out := false;
AA: if (x+2.01×H—xE>0)≡(H>0) then
    begin Hs := H; out := true; H := (xE—x)/2 end if;
    RK1ST(x, y, 2×H, x1, y1);

BB: RK1ST(x, y, H, x2, y2); RK1ST(x2, y2, H, x3, y3);
    for k := 1 step 1 until n do
        if comp(y1[k], y3[k], eta) > eps then goto CC;
    comment comp (a, b, c) is a function designator, the value of which
    is the absolute value of the difference of the mantissae of a and b, after
    the exponents of these quantities have been made equal to the largest of
    the exponents of the originally given parameters a, b, c;
    x := x3; if out then goto DD;
    for k := 1 step 1 until n do y[k] := y3[k];
    if s=5 then begin s := 0; H := 2×H end if;
    s := s+1; goto AA;
```

CC: $H := 0.5 \times H$; *out* := **false**; $x1 = x2$;
    for $k := 1$ **step** 1 **until** n **do** $y1[k] := y2[k]$;
    goto BB;

DD: **for** $k := 1$ **step** 1 **until** n **do** $yE[k] := y3[k]$
end RK

Alphabetic index of definitions of concepts and syntactic units

All references are given through section numbers. The references are given in three groups:

def   Following the abbreviation "def", reference to the syntactic definition (if any) is given.

synt Following the abbreviation "synt", references to the occurrences in metalinguistic formulae are given. References already quoted in the def-group are not repeated.

text Following the word "text", the references to definitions given in the text are given.

The basic symbols represented by signs other than underlined *(bold faced. Publisher's remark)* words have been collected at the beginning. The examples have been ignored in compiling the index.

$+$ see: plus
$-$ see: minus
$\times$ see: multiply
$/ \div$ see: divide
$\uparrow$ see: exponentiation
$< \leqq = \geqq > \neq$ see: ⟨relational operator⟩
$\equiv \supset \vee \wedge \neg$ see: ⟨logical operator⟩
,   see: comma
.   see: decimal point
$_{10}$ see: ten
:   see: colon
;   see: semicolon
:= see: colon equal
⊔   see: space
() see: parentheses
[] see: subscript bracket
''   see: string quote

⟨actual parameter⟩, def 3.2.1, 4.7.1
⟨actual parameter list⟩, def 3.2.1, 4.7.1
⟨actual parameter part⟩, def 3.2.1, 4.7.1
⟨adding operator⟩, def 3.3.1

Note. This report is published in *Numerische Mathematik,* in the *Communications of the ACM,* and in the *Journal of the British Computer Soc.* Reproduction of this report for any purpose is explicitly permitted; reference should be made to this issue of *Numerische Mathematik* and to the respective issues of the *Communications* and the *Journal of the British Computer Soc.* as the source.

Technical University Delft

Delft, Holland

W. L. van der Poel,

(Chairman of Working Group 2.1 on ALGOL of the

International Federation for Information Processing)

Report on SUBSET ALGOL 60 (IFIP)[1]

Introduction

The present report contains the results of the work of IFIP Working Group 2.1 (WG 2.1) on establishing a subset of the algorithmic language ALGOL 60 as defined in the Revised Report on the Algorithmic Language ALGOL 60 [Numerische Mathematik **4**, 420 (1963), Communications of the ACM **6**, 1 (1963), and The Computer Journal **5**, 349 (1963)].

The meaning of a subset of ALGOL 60

By a subset of ALGOL 60 is here meant a language such that every program written in the subset language is automatically also a program written in ALGOL 60 and has the same meaning in both languages.

The purposes of establishing a subset

The main incentive to the work is the realization that the generality of some of the features of ALGOL 60 and the disagreement concerning the exact meaning of others have proved a considerable discouragement to some of the groups who have contemplated implementing the language and have caused most of the existing implementations to be based on subsets defined locally. Clearly, if no attempt is made to avoid it, this development will lead to the use of any number of languages, all subsets of ALGOL 60, but in many cases mutually incompatible ones.

This state of affairs means a weakening of the efforts of the authors of ALGOL 60 toward establishing a common language, which may still be avoided if a subset which avoids the above-mentioned difficulties were defined and recommended by the official working group.

In lending its support to this development the IFIP/WG 2.1 is not blind to the fact that to a certain extent it may reduce the effort spent on implementing the full ALGOL 60 language, and thus may decrease the effect of this language on users and implementors. However, this negative effect is counteracted not only by the positive benefits of the subset already mentioned, but also by the fact that the impact of the full ALGOL 60 on the computing world has already been very considerable, as evidenced by the attention given to it at recent meetings and in the literature.

---

[1] Numerische Mathematik **6**, 454—458 (1964).

The development of the subset

The members of WG 2.1 wish to acknowledge the extensive and valuable work on the establishment of subsets done by other bodies. The present SUBSET ALGOL 60 is, in fact, to a large extent based on such work, as is evident from the following historical notes.

The work of WG 2.1 on defining the subset was begun during a meeting in Munich in August 1962. During this meeting a detailed comparison of two existing subsets, the ALCOR subset [ALGOL-Manual der ALCOR-Gruppe. Elektronische Rechenanlagen, **3**, 206 (1961)], and SMALGOL [Smalgol-61. Communications of the ACM, **4**, 499 (1961)], was made, and tentative decisions on a number of the characteristics of the subset were taken. By September 1963, when the next meeting of WG 2.1 took place in Delft, a proposal for a subset prepared by the European Computer Manufacturers Association (ECMA) was kindly made available to WG 2.1. This proved to be particularly valuable to the WG 2.1 because of the care with which its stipulations had been phrased and was in fact used verbatim for many of the definitions of the present subset.

The form of the definition of the subset

The description of the subset consists of (1) the corrections necessary to convert the Revised Report on the Algorithmic Language ALGOL 60 into a report defining the SUBSET ALGOL 60, and (2) explanatory remarks describing the effect of the defining changes in an informal manner only.

Definition of SUBSET ALGOL 60 in terms of the Revised ALGOL 60 Report.

Section	Subset definitions	Explanation
2.1.	Delete: "$\|A\|B\| \ldots \|Y\|Z$". Delete: ", or extended … delimiter)". Add: "Note: If a particular implementation requires capitals rather than small letters, one must regard them as a hardware representation for the small letters".	Only one case of letters is provided for.
2.3.	Delete: "$\| \div$"	The so-called integer division is not included in the subset.
2.3.	Delete from definition of declarator "**own**$\|$".[1]	The **own** concept is not included in the subset.

---

[1] [Where typewritten copy uses underlining, this is replaced by boldface type in printed copy. — Ed.]

2.4.3.   Replace: "They may be chosen freely" by: "Identifiers may be chosen freely; but the effects due to the occurrence of two different identifiers the first six basic symbols of which are common are undefined".

In the subset identifiers are differentiated only up to six leading basic symbols.

3.3.1.   Delete: "| ÷"

3.3.4.   Replace the words: "the following rules" of the last sentence by: "a set of rules. However, if the type of an arithmetic expression according to the rules cannot be determined without evaluating an expression or ascertaining the type or value of an actual parameter, it is **real**. These rules are".

In the subset the type of an arithmetic expression will be in certain cases **real** where it will be **integer** in ALGOL 60. Thus arithmetic will be less precise in some cases.

3.3.4.2.   Replace: "The operations ... both denote" by: "The operation $\langle\text{term}\rangle / \langle\text{factor}\rangle$ denotes". Delete last sentence.

3.3.4.3.   Insert between "... rules" and ":": "with the exception that, if both the basis a and the exponent i are of **integer** type, then the exponent has to be an unsigned integer, otherwise the result is undefined".

Exponentiation with integer basis and exponent is restricted in the subset.

3.3.5.1.   Delete: " ÷ "

3.5.1.   Delete "|⟨unsigned integer⟩".

Integer labels are not provided for.

3.5.1.   Replace the last two formulae by: "⟨designational expression⟩ ::= ⟨label⟩|⟨switch designator⟩".

In the subset only unconditional and unparenthesized designational expressions are provided for.

3.5.3.   Delete: "In the general case ... is already found.". Replace "selects one of the designational expressions ... a recursive process" by: "selects one of the labels con-

See 5.3.1.

tained in the switch list of the switch declaration. The selection is obtained by counting these labels from left to right".

3.5.5. Delete.

4.3.5. Delete.

The effect of a **goto** statement involving an undefined switch designator is undefined in the subset.

4.6.1. Replace: "⟨for clause⟩ ... **do**" by: "⟨for clause⟩ ::= **for** ⟨variable identifier⟩ := ⟨for list⟩ **do**".

The controlled variable in a for clause is restricted in the subset to be a variable identifier.

4.7.3.2. Replace: "after enclosing this ... syntactically possible" by: "this actual parameter being an identifier, or string, otherwise the name replacement is undefined".

In name replacement (call by name) the actual parameter can only be an identifier or a string.

4.7.5. Insert after: "... ALGOL statement" and before ".": "in the sense of this subset".

4.7.5.5. Replace by: "Kind and type of actual parameters must be the same as those of the corresponding formal parameters, if called by name".

4.7.5. Add section 4.7.5.6: "No call of the procedure itself may occur during the execution of the statements of the body of any procedure, nor during the evaluation of those of its actual parameters, the corresponding formal parameters of which are called by name, nor during the evaluation of expressions occurring in declarations inside the procedure".

Recursive procedures and recursive use of procedures are not included.

5. Delete first two sentences of fourth paragraph.

5.   Insert after "... any one block
        head." and before "Syntax":
        "The identifier associated with a
        quantity declared in a declaration
        may not occur denoting that
        quantity more than once between
        the **begin** of the block in whose
        head that declaration occurs and
        the semicolon which ends that de-
        claration, excepting the case where
        this occurrence is the occurrence
        of a procedure identifier in the left
        part list of an assignment state-
        ment in the sense of section 5.4.4.".

5.1.1. Replace the last two metalinguis-
        tic formulae by:
        "⟨type declaration⟩ ::=
            ⟨type⟩ ⟨type list⟩".

5.1.3. Delete last sentence.

5.2.1. Replace the last formula by:
        "⟨array declaration⟩ ::=
            array ⟨array list⟩|
            ⟨type⟩ **array** ⟨array list⟩".

5.2.5. Delete "even if an array is de-
        clared **own**".

5.3.1. Replace: "⟨switch list⟩ ::=
            ⟨designational expression⟩|
            ⟨switch list⟩, ⟨designational
            expression⟩" by:
        "⟨switch list⟩ ::= ⟨label⟩|
            ⟨switch list⟩, ⟨label⟩".

        In the subset the desig-
        national expressions in a
        switch list are restricted to
        be labels only.

5.3.3. Replace: "These values ... its asso-
        ciated integer" by: "These values
        are given as labels entered in the
        switch list. With each of these
        labels there is associated a positive
        integer 1, 2, ..., obtained by count-
        ing the items in the list from left
        to right. The value of the switch
        designator corresponding to a
        given value of the subscript ex-

20*

pression (cf. section 3.5. Designational expressions) is the label in the switch list having this given value as its associated integer".

5.3.4. Delete.

5.3.5. Replace by: "Influence of scopes. If a switch designator occurs outside the scope of a label in the switch list, and an evaluation of this switch designator selects this label, then a possible conflict between the identifier used to denote this label and an identifier whose declaration is valid at the place of the switch designator will be avoided by a suitable change of this latter identifier.".

5.4.3. Add: "No identifier may occur more than once in a formal parameter list.".

5.4.4. Delete last sentence.

5.4.4. Add to text: "A function designator must be such that all its possible uses in the form of a procedure statement are equivalent to dummy statements".

5.4.5. Replace third sentence by: "Specifications of all formal parameters if any must be supplied". Complete specification parts are required.

This report has been reviewed by IFIP/TC 2 on Programming Languages in May 1964 and has been approved by the Council of the International Federation for Information Processing. Reproduction of this report for any purpose is explicitly permitted only in full. In making reference to this report the name IFIP SUBSET ALGOL 60 must be mentioned. IFIP does not authorise the language described in this report to be referred to as ALGOL without adding the word SUBSET.

International Federation for Information Processing
Working Group 2.1 on ALGOL
Chairman: W. L. VAN DER POEL,
Technological University Delft,
Delft, Netherlands

Report on Input-Output Procedures for ALGOL 60[1]

1. Introduction

It was recognized in the IFIP/WG 2.1 on ALGOL that some procedures to be used in connection with input and output are considered as being primitives, which cannot be expressed otherwise than by means of a code body. Among these are the following ones:

$$
\begin{aligned}
&insymbol\\
&outsymbol\\
&length\\
&inreal\\
&outreal\\
&inarray\\
&outarray.
\end{aligned}
\qquad (1)
$$

Apart from these primitives one needs in practice a fuller set of input-output procedures. However, the language ALGOL 60 is so flexible that different schemes of I/O procedures can be defined in it largely by means of the primitives mentioned above. A few examples of this will be given in section 4 of this report.

2. Definitions

It is recommended that, if not otherwise declared, the identifiers (1) will be associated with procedures which transfer values between the variables of the program and values carried in any kind of foreign media not otherwise accessible from the program.

The corresponding procedure declarations are:

procedure *insymbol* (*channel, string, destination*); **value** *channel*;
　　　　integer *channel, destination*; **string** *string*;
　　　　⟨procedure body⟩

procedure *outsymbol* (*channel, string, source*); **value** *channel, source*;
　　　　integer *channel, source*; **string** *string*; ⟨procedure body⟩

integer procedure *length* (*string*); **string** *string*; ⟨procedure body⟩

procedure *inreal* (*channel, destination*); **value** *channel*;
　　　　integer *channel*; **real** *destination*; ⟨procedure body⟩

[1] Numerische Mathematik **6**, 459—462 (1964).

procedure *outreal (channel, source)*; **value** *channel, source*;
        integer *channel*; **real** *source*; ⟨procedure body⟩

procedure *inarray (channel, destination)*; **value** *channel*;
        integer *channel*; **array** *destination*; ⟨procedure body⟩

procedure *outarray (channel, source)*; **value** *channel*;
        integer *channel*; **array** *source*; ⟨procedure body⟩

The procedure statements and the function designator calling these procedures must have the following forms:

insymbol (⟨arithmetic expression⟩ ⟨parameter delimiter⟩ ⟨string⟩
       ⟨parameter delimiter⟩ ⟨variable⟩)
outsymbol (⟨arithmetic expression⟩ ⟨parameter delimiter⟩ ⟨string⟩
       ⟨parameter delimiter⟩ ⟨arithmetic expression⟩)
length (⟨string⟩)
inreal (⟨arithmetic expression⟩ ⟨parameter delimiter⟩ ⟨variable⟩)
outreal (⟨arithmetic expression⟩ ⟨parameter delimiter⟩
       ⟨arithmetic expression⟩)
inarray (⟨arithmetic expression⟩ ⟨parameter delimiter⟩
       ⟨array identifier⟩)
outarray (⟨arithmetic expression⟩ ⟨parameter delimiter⟩
       ⟨array identifier⟩)

In all these cases, except for the call of length, the value of the first actual parameter must be a positive integer identifying an input or output channel available to the program.

3. Actions of the procedure bodies

The pair of procedures *insymbol* and *outsymbol* provides the means of communicating between foreign media and the variables of the program in terms of single basic symbols or any additional symbols. In either procedure the correspondence between the basic symbols and the values of variables in the program is established by mapping the sequence of the basic symbols given in the string supplied as the second parameter, taken in the order from left to right, onto the positive integers 1, 2, 3, ... Using this correspondence the procedure *insymbol* will assign to the type **integer** variable given as the third parameter the value corresponding to the next basic symbol appearing on the foreign medium. If this next basic symbol does not appear in the string given as the second parameter, the number 0 will be assigned. If the next symbol appearing in the input is not a basic symbol of ALGOL 60 a negative integer, corresponding to the symbol, will be assigned.

Similarly the procedure *outsymbol* will transfer the basic symbol corresponding to the value of the third parameter to the foreign medium.

If the value of the third parameter is negative a symbol corresponding to this value will be transferred. It is understood that where the foreign medium may be used both for *insymbol* and *outsymbol*, the negative integer values associated with each additional symbol will be the same for the two procedures. More generally, if additional symbols are used the corresponding values must be given as accompanying information with the program (cf. the footnote to section 1 of the Revised ALGOL 60 Report).

The type procedure *length* is introduced to enable the calculation of the length of a given (actual or formal) string to be made (cf. example *outstring*). The value of *length(s)* is equal to the number of basic symbols of the open string enclosed between the outermost string quotes.

The two procedures *inreal* and *outreal* form a pair. The procedure *inreal* will assign the next value appearing on the foreign medium to the **real** type variable given as the second parameter. Similarly, procedure *outreal* will transfer the value of the second actual parameter to the foreign medium.

The representation of values on the foreign media will not be further described, except that it is understood that in so far as a medium can be used for both input and output a value which has been transferred to a given medium with the aid of a call of *outreal* will be represented in such a way that the same value, in the sense of numerical analysis (cf. section 3.3.6), may be transferred back to a variable by means of procedure *inreal*, provided that an appropriate manipulation of the foreign medium has also been performed.

Procedures *inarray* and *outarray* also form a pair; they transfer the ordered set of numbers forming the value of the array given as the second parameter, the array bounds being defined by the corresponding array declaration rather than by additional parameters (the mechanism for doing that is already available in ALGOL 60 for the value call of arrays). The order in which the elements of the array are transferred corresponds to the lexicographic order of the values of the subscripts, i.e.

$$a[k_1, k_2, \ldots, k_m] \quad \text{precedes} \quad a[j_1, j_2, \ldots, j_m]$$

$$\left. \begin{array}{l} \text{provided } k_i = j_i \quad (i = 1, 2, \ldots, p-1) \\ k_p < j_p \end{array} \right\} \quad (1 \leq p \leq m). \tag{2}$$

It should be recognized that the possibly multidimensional structure of the array is not reflected in the corresponding numbers on the foreign medium, where they appear only as a linear sequence as defined by (2).

The representation of the numbers on the foreign medium conforms to the same rules as given for *inreal* and *outreal*; in fact it is possible for example to input numbers by *inreal* which before have been output by *outarray*.

4. Examples

procedure *outboolean* (*channel, boolean*); **value** *boolean*; **integer** *channel*;
  Boolean *boolean*; **comment** this procedure outputs a **Boolean** value
  as a basic symbol **true** or **false**;
  if *boolean* **then** *outsymbol* (*channel*, '**true**', 1)
            else *outsymbol* (*channel*, '**false**', 1)

procedure *outstring* (*channel, string*); **value** *channel*; **integer** *channel*;
  string *string*; **comment** outputs the **string** *string* to the foreign
                      medium;
  begin integer *i*;
        for *i* := 1 **step** 1 **until** *length* (*string*) **do** *outsymbol* (*channel*,
                string, i)
  end

procedure *ininteger* (*channel, integer*); **value** *channel*;
  integer *channel, integer*;
  comment inputs an integer which on the foreign medium appears
  as a sequence of digits, possibly preceded by a sign, and followed by
  a comma. Any other symbol in front of the sign is discarded;
  begin integer *n, k*; **Boolean** *b*;
        integer := 0; *b* := **true**;
        for *k* := 1, *k* + 1 **while** *n* = 0
                    do *insymbol*(*channel*, '0123456789 − +', *n*);
        if *n* = 11 **then** *b* := **false**; **if** *n* > 10 **then** *n* := 1;
        for *k* := 1, *k* + 1 **while** *n* ≠ 13 **do**
        begin *integer* := 10 × *integer* + *n* − 1;
              insymbol (*channel*, '0123456789 − +,', *n*)
        end 1;
        if ¬ *b* **then** *integer* := − *integer*
  end
begin
      begin array *a*[1:10];
            ⟨statements⟩;
            outarray (15, *a*)
      end;
      begin array *b*[0:1, 1:5];
            inarray (15, *b*);
            ⟨statements⟩
      end
end

The following example exhibits the use of *inarray* and *outarray* for
inversion of a matrix including transfer of the matrix elements from and

to the foreign medium. It requires that an appropriate declaration for a matrix inversion procedure as well as the declaration of *outstring* as given above are inserted at appropriate places in the program.

begin integer n;
      inreal$(5, n)$; **comment** the matrix elements must be preceded by
                                         the order;
      begin array $a[1:n, 1:n]$;
          inarray$(5, a)$;
          matrix inversion$(n, a, singular)$;
          outarray$(15, a)$;
          goto *ex*
      end;
singular: *outstring*$(15, \text{'}singular\text{'})$;
ex: **end**

5. Concluding remarks

WG 2.1 does not propose any further means for input-output operations, but would like to draw attention to

"A proposal for input-output conventions in ALGOL 60", by the ACM programming languages committee (Subcommittee on ALGOL, D. E. KNUTH, chairman), and to the extensive list of references at the end of that report.

This report has been reviewed by IFIP/TC 2 on Programming Languages in May 1964 and has been approved by the Council of the International Federation for Information Processing. Reproduction of this report for any purpose is explicitly permitted only in full.

International Federation for Information Processing
Working Group 2.1 on ALGOL
Chairman: W. L. VAN DER POEL
Technological University Delft
Delft, Netherlands

References

[1] ADAMS, C. W., and J. H. LANING jr.: The MIT systems of automatic coding. Proceedings of a Symposium on Automatic Programming for Digital Computers, p. 40—68. Washington DC, May 1954.

[2] Algol-Bulletin. Aperiodically issued publication. No 1—15 (Nov. 1959—June 1962) ed. by P. NAUR, No 16 and following (May 1964 and later) ed. by F. DUNCAN.

[3] ARNOLDI, E. W.: The principle of minimized iterations in the solution of eigenvalue problems. Quart. Appl. Math. 9, 17—29 (1951).

[4] BACHELOR, G. A., J. R. H. DEMPSTER, D. E. KNUTH, and J. SPERONI: Smalgol 61. Comm. Ass. Comp. Mach. 4, 499—502 (1961).

[5] BACKUS, J. W., F. L. BAUER, H. BOTTENBRUCH, C. KATZ, A. J. PERLIS (Ed.), H. RUTISHAUSER, K. SAMELSON (Ed.), and J. H. WEGSTEIN:
(a) Report on the algorithmic language ALGOL. Num. Math. 1, 41—60 (1959).
(b) Preliminary Report-Internat. Algebraic Language. Comm. Ass. Comp. Mach. 1, No 12, 8—22 (1958) (these are two identical reports).

[6] — — J. GREEN, C. KATZ, J. McCARTHY, P. NAUR (Ed.), A. J. PERILS, H. RUTISHAUSER, K. SAMELSON, B. VAUQUOIS, J. H. WEGSTEIN, A. VAN WIJNGAARDEN, and M. WOODGER: Report on the Algorithmic Language ALGOL 60.
(a) Num. Math. 2, 106—136 (1960).
(b) Com. Ass. Comp. Mach. 3, No 5, 299—314 (1960).

[7] — — — — — — — — — — Revised Report on the Algorithmic Language ALGOL 60.
(a) Num. Math. 4, 420—453 (1962/63).
(b) Comm. Ass. Comp. Mach. 6, No 1, 1—17 (1963).
(c) Computer J. 5, 349—367 (1962/63).
See also this volume, Appendix B.

[8] BAUER, F. L., H. RUTISHAUSER, and E. STIEFEL: New aspects in numerical quadrature. Proceedings of Symposia in Applied Mathematics, vol. XV. Amer. Math. Soc. Providence, RI, 1963, p. 199—218.

[9] BAUMANN, R.: ALGOL-Manual der ALCOR-Gruppe. Elektronische Rechenanlagen 3, 206—212, 259—265 (1961); 4, 71—85 (1962).

[10] BOEHM, C.: Calculatrices digitales du dechiffrage de formules logico-mathematiques par la machine meme. Thesis ETH Zürich 1954.

[11] Control Data Corporation, CODAP1 Reference Manual. Control Data Corporation, Minneapolis 1963.

[12] Deutscher Normenausschuß, Darstellung von ALGOL-Symbolen auf 5-Spur-Lochstreifen und 80-spaltigen Lochkarten. DIN-Normblatt 66006, Beuth Vertrieb GMBH, Berlin 1965.

[13] ERSHOV, A. P.: Programming programme for the BESM computer (translated from Russian by M. NADLER), 158 p. London: Pergamon Press 1959.

[14] FORSYTHE, G. E.: Generation and use of orthogonal polynomials for data fitting. J. Soc. Ind. Appl. Math. 5, 74—78 (1957).

[15] GLENNIE, A. E.: Unpublished report.

[16] GRAU, A. A.: On the reduction of number range in the use of the Graeffe process. J. Ass. Comp. Mach. **10**, 538—544 (1963).

[17] HENRICI, P.: Discrete variable methods in ordinary differential equations, 407 p. New York and London: John Wiley & Sons 1962.

[18] HOFFMANN, W., u. A. WALTHER: Elektronische Rechenmaschinen und Informationsverarbeitung. Nachrichtentechnische Fachberichte, Bd. 4. 229 S. Braunschweig: F. Vieweg & Sohn 1956.

[19] IBM, Specifications for the IBM mathematical formula translating system FORTRAN. Preliminary report. Applied Science Division, Internat. Business Machines Corporation, New York 1954.

[20] IFIP, Report on SUBSET ALGOL 60. Ed. by W. L. VAN DER POEL.
(a) Num. Math. **6**, 454—458 (1964).
(b) Comm. Ass. Comp. Mach. **7**, 626—628 (1964).
See also this volume, Appendix B.

[21] IFIP, Report on Input-Output-Procedures for ALGOL 60.
(a) Num. Math. **6**, 459—462 (1964).
(b) Comm. Ass. Comp. Mach. **7**, 628—630 (1964).
See also this volume, Appendix B.

[22] KNUTH, D.: Proposal for input-output conventions in ALGOL 60. Report of the Subcommittee of the ACM Programming Languages Committee. Comm. Ass. Comp. Mach. **7**, 273—283 (1963).

[23] KUNTZMANN, J.: Methodes numériques. Interpolations-dérivées. 252 p. Paris: Dunod 1959.

[24] NEVILLE, E. H.: Iterative interpolation. J. Indian Math. Soc. **20**, 87—120 (1934).

[25] NYSTROEM, E. J.: Über die numerische Integration von Differentialgleichungen. Acta Soc. Sci. Fenn. **50**, No 13, 1—55 (1925).

[26] RUTISHAUSER, H.: Automatische Rechenplanfertigung bei programmgesteuerten Rechenmaschinen. Mitt. Nr 3 aus dem Inst. für angewandte Math. der ETH, 45 p. Basel: Birkhäuser 1952.

[27] — Bemerkungen zum programmgesteuerten Rechnen. Vorträge über Rechenanlagen, S. 34—37. Max-Planck-Institut für Physik, Göttingen 1953.

[28] — Der Quotienten-Differenzen-Algorithmus. Mitt. Nr 7 aus dem Inst. für angew. Math. der ETH, 74 p. Basel: Birkhäuser 1957.

[29] — Zur Matrizeninversion nach Gauß-Jordan. Z. angew. Math. u. Phys. **10**, 281—291 (1959).

[30] — Interference with an ALGOL procedure. Annual Review in Automatic Programming 2, 67—75. London: Pergamon Press 1961.

[31] — Stabile Sonderfälle des Quotienten-Differenzen-Algorithmus. Num. Math. **5**, 95—112 (1963).

[32] SAMELSON, K., u. F. L. BAUER: Sequentielle Formelübersetzung. Elektronische Rechenanlagen 1, 176—182 (1959).

[33] —, and F. L. BAUER: Sequential formula translation. Comm. Ass. Comp. Mach. **3**, No 2, 76—83 (1960).

[34] SCHWARZ, H. R.: Introduction to ALGOL. Comm. Ass. Comp. Mach. **5**, 82—95 (1962).

[35] STIEFEL, E.: Kernel polynomials and their applications. NBS-AMS-Series **49**, 1—23. Washington D.C.: National Bureau of Standards 1958.

[36] STOCK, J. R.: Die mathematischen Grundlagen für die Organisation der elektronischen Rechenmaschine der ETH. Mitt. Nr 6 aus dem Inst. für angew. Math. der ETH, 76 p. Basel: Birkhäuser 1956.

[37] TAYLOR, W. J.: Method of Lagrangian curvilinear interpolation. J. Research
     Nat. Bur. Standards **35**, 151—155 (1945).
[38] THURNAU, D. H.: Algorithm 195. Comm. Ass. Comp. Mach. **6**, 441 (1963).
[39] WALL, H. S.: Analytic theory of continued fractions. Princeton: Van Nostrand
     1948.
[40] WOODGER, M.: Supplement to the ALGOL-report. Comm. Ass. Comp. Mach.
     6, 18—20 (1963).
[41] WYNN, P.: On a device for computing the $e_m(S_n)$ transformation. Math. Tables
     and Other Aids to Comp. **10**, 91—96 (1956).
[42] — Upon systems of recursion which obtain among the quotients of the Pade
     table. To appear.
[43] ZURMUEHL, R.: Praktische Mathematik für Ingenieure und Physiker, 3. Aufl.,
     548 S. Berlin-Göttingen-Heidelberg: Springer 1961.
[44] — Matrizen, 4. Aufl., 452 S. Berlin-Göttingen-Heidelberg: Springer 1964.
[45] ZUSE, K.: Über den allgemeinen Plankalkül als Mittel zur Formulierung sche-
     matisch-kombinativer Aufgaben. Arch. Math. **1**, 441—449 (1948/49).

Books on ALGOL

[46] AGEEV, M. I.: The fundamentals of the algorithmic language ALGOL 60 [Rus-
     sian]. Obschichie Voprosy Programmirovaniya 1, 116 p. Vychisl. Tsentr
     AN SSSR, 1964.
[47] ANDERSEN, C.: Introduction to ALGOL 60, 57 p. Addison Wesley, Reading
     1964.
[48] ARSAC, J., L. A. LENTIN, M. NIVAT et L. NOLIN: ALGOL 60. Théorie et prati-
     que, 203 p. Paris: Gauthier-Villars 1965.
[49] BAUMANN, R.: ALGOL-Manual der ALCOR-Gruppe, 176 S. München u. Wien:
     Oldenbourg 1965.
[50] — M. FELICIANO, F. L. BAUER, and K. SAMELSON: Introduction to ALGOL 60,
     142 p. Englewood Cliffs: Prentice-Hall 1964.
[51] BOLLIET, L., N. GASTINEL et P. J. LAURENT: Un nouveau language scientifi-
     que ALGOL, manuel pratique, 196 p. Paris: Hermann 1964.
[52] DIJKSTRA, E. W.: A primer of ALGOL 60 programming, 114 p. New York:
     Academic Press 1962.
[53] EKMAN, T., and C. E. FRÖBERG: Introduction to ALGOL programming, 123 p.
     Lund: Studentlitteratur 1965.
[54] KERNER, I.O., u. G. ZIELKE: Einführung in die algorithmische Sprache ALGOL,
     283 S. Leipzig: Teubner 1966.
[55] NICKEL, K.: ALGOL-Praktikum, 220 S. Karlsruhe: G. Braun 1964.
[56] REEVES, C. M., and M. WELLS: A course on programming in ALGOL 60, 82 p.
     London: Chapman & Hall 1964.
[57] THURNAU, D. H., R. E. JOHNSON, and R. J. HAM: ALGOL-programming — a
     basic approach, 158 p. Denver: Big Mountain Press 1964.

Subject Index

Die Grundlehren der mathematischen Wissenschaften in Einzeldarstellungen mit besonderer Berücksichtigung der Anwendungsgebiete